Materials of
CONSTRUCTION

MAXWELL PRESS

Materials of CONSTRUCTION

Harry E. Pulver B.S., C.E

Associate Professor of Civil and Structural Engineering The University of Wisconsin

Chennai Trichy New Delhi

MAXWELL PRESS

This edition has been published in india by arrangement with Carson Books, UK

ISBN 978-93-5528-222-4

Maxwell Press
No. 44, Nallathambi Street,
Triplicane, Chennai 600 005

Printed and bound in India

MJP 1292

Publisher : C. Janarthanan

Publisher's Note

The legacy of a country is in its varied cultural heritage, historical literature, developments in the field of economy and science. The top nations in the world are competing in the field of science, economy and literature. This vast legacy has to be conserved and documented so that it can be bestowed to the future generation. The knowledge of this legacy is slowly getting perished in the present generation due to lack of documentation.

Keeping this in mind, the concern with retrospective acquiring of rare books has been accented recently by the burgeoning reprint industry. Maxwell Press is gratified to retrieve the rare collections with a view to bring back those books that were landmarks in their time.

In this effort, a series of rare books would be republished under the banner, "Maxwell Press". The books in the reprint series have been carefully selected for their contemporary usefulness as well as their historical importance within the intellectual. We reconstruct the book with slight enhancements made for better presentation, without affecting the contents of the original edition.

Most of the works selected for republishing covers a huge range of subjects, from history to anthropology. We believe this reprint edition will be a service to the numerous researchers and practitioners active in this fascinating field. We allow readers to experience the wonder of peering into a scholarly work of the highest order and seminal significance.

MAXWELL PRESS

PREFACE

This textbook has been prepared for use in a correspondence-study course offered by the Extension Division of The University of Wisconsin. It is intended that the book will be supplemented by questions and problems and such extra material as is found to be advisable. The book is quite elementary and is suitable for students who have had an ordinary training in English and Arithmetic.

It is believed that this text may also be of use for residence courses in technical schools, as the book is of such size that its entire contents may be covered in the length of time usually assigned to this subject. The text should be used in connection with courses in the Strength of Materials and Materials Testing as no attempt has been made to cover any of the subject matter usually included in those courses, with the exception of a few tests and specification requirements.

The data in this book have been compiled from many sources and the author has endeavored to give credit where it is due.

The author is indebted to Professor D. A. Abrams of Lewis Institute for material on concrete proportioning and to the several manufacturing companies and individuals for illustrations furnished.

H. E. Pulver.

The University of Wisconsin,
Madison, Wisconsin.
June, 1922.

CONTENTS

CHAPTER III. PORTLAND CEMENT

A. Definition and Classification

B. Manufacture of Portland Cement

C. Properties and Uses of Portland Cement

CHAPTER IV. PORTLAND CEMENT MORTARS

A. Definitions and Materials

B. Proportioning and Mixing Mortar

C. Properties of Portland Cement Mortars

CHAPTER V. PLAIN CONCRETE

A. Definitions and Materials

B. Proportioning of Concrete

C. Mixing of Concrete

D. Deposition of Concrete

E. Impervious Concrete

C. PROPERTIES OF BUILDING STONE

CHAPTER VII. BRICK AND OTHER CLAY PRODUCTS

A. CLASSIFICATIONS AND DEFINITIONS

B. MANUFACTURE OF CLAY BUILDING BRICK

C. MANUFACTURE OF OTHER BRICK

D. OTHER CLAY PRODUCTS

E. PROPERTIES OF BRICK AND OTHER CLAY PRODUCTS

CHAPTER VIII. STONE AND BRICK MASONRY

A. STONE MASONRY

B. Brick and Hollow Tile Masonry

CHAPTER IX. TIMBER

A. Trees

B. Preparing the Timber

F. MALLEABLE CAST IRON

CHAPTER XII. WROUGHT IRON

A. DEFINITION AND CLASSIFICATIONS

B. MANUFACTURE OF WROUGHT IRON

C. CONSTITUTION, PROPERTIES, AND USES OF WROUGHT IRON

CHAPTER XIII. STEEL

A. DEFINITIONS AND CLASSIFICATIONS

B. METHODS OF MANUFACTURE OF STEEL

MATERIALS OF CONSTRUCTION

CHAPTER I

PLASTERS AND NATURAL CEMENTS

A. GYPSUM PLASTERS

1. Definition and Classification.—Gypsum plasters may be defined as those plasters which are produced by the partial or complete dehydration of gypsum. Pure gypsum is a mixture of 1 part of calcium sulphate ($CaSO_4$) and 2 parts of water ($2H_2O$). Gypsum plasters may be classified as follows:

1. Those produced by the incomplete dehydration of the gypsum, the calcination being carried on at a temperature less than 400 degrees Fahrenheit.
   - (a) Plaster of Paris ($CaSO_4 + \frac{1}{2}H_2O$), in which no foreign material has been added either during or after the calcination.
   - (b) Cement plaster, which is made from an impure gypsum or by adding certain impurities, during the manufacture, to act as a retarder to the plaster. This plaster is often called hard wall or patent plaster.
2. Those plasters produced by the complete dehydration of the gypsum, the calcination being carried on at temperatures greater than 400 degrees Fahrenheit.
   - (a) Calcined plaster (flooring plaster), a pure calcined gypsum.
   - (b) Hard-finish plaster, which is made by calcining gypsum at a red heat or higher temperature and to which certain substances, such as alum or borax, have been added.

2. Manufacture of Gypsum Plasters.—The process of manufacture is essentially the same for all of the first class of gypsum plasters, variations being made in the temperature of calcination and the purity of the gypsum used. The raw material is a natural gypsum rock usually containing from 1 to 6 per cent of impurities. The rock is crushed, ground to a powder, and then heated in a large calcining kettle. If a rotary calciner is used, the fine grinding is done after the calcination. In making plaster of Paris, a pure gypsum is calcined at a temperature of about 220 degrees Fahrenheit thus driving off three-fourths of the water present. Cement plaster is made in the same way, an impure gypsum being used. Often a retarder (a substance to make the plaster slow setting) is added after the calcination.

Flooring plaster is made by calcining lumps of gypsum in a separate feed kiln similar to the kiln used for the calcination of lime. The temperature of calcination is usually about 850 degrees Fahrenheit and the time required is about 3 hours. Higher temperatures or longer heating will burn the plaster and

FIG. 1.—Rotary cylinder type of plaster calciner.

cause it to lose its powers of setting and hardening. After the calcination, the plaster must be finely ground.

Keene's cement is the best-known kind of hard-finish plaster. This plaster is made by calcining a very pure gypsum at a red heat, immersing it in a 10 per cent alum solution, recalcining it, and then finely grinding the calcined material.

3. Properties of Gypsum Plasters.—All gypsum plasters will set or harden when mixed with the proper amount of water. The process is a combination of the plaster and the water to form gypsum. The time required varies from 5 minutes to 2 hours according to the plaster used and the conditions of the mixing. The plasters made from pure gypsum are the quicker setting while the hard-burned plasters form the harder substances.

Very little data are available concerning the strength of plasters, and the methods and conditions of testing have never been standardized. The strength varies according to the quality of the plaster, the quality of the sand, and the care taken in the mixing. The following table will give an idea of the strength of good plasters:

TENSION TEST

Neat paste..........	Age 1 month	Strength, about 350 lb. per square inch
1:2 sand mortar.....	Age 1 month	Strength, about 175 lb. per square inch

COMPRESSION TEST

Neat paste.......	Age 1 month	Strength, 1,200 to 2,000 lb. per square inch
1:2 sand mortar..	Age 1 month	Strength, 900 to 1,500 lb. per square inch

ADHESION TEST

Plaster to paving brick..............	Age 1 month	Strength, about 100 lb. per square inch
Plaster to 1:2 mortar.	Age 1 month	Strength, about 130 lb. per square inch

The results of tests have shown that neat plasters gain rapidly in strength for the first few days only, and that the maximum tensile and compressive strength of the plasters is reached in from 2 to 4 weeks. The mortars gain in strength a little less rapidly than the neat plasters.

4. Uses of Gypsum Plasters.—Gypsum plasters are not used very much as a material for engineering construction. Plaster of Paris is used as a casting plaster and for making quick repairs, etc., where a quick setting plaster is desired. Wall plasters are made by adding lime and a retarder, together with hair, wood fiber, etc., to a calcined plaster. (Ordinary wall plaster contains no gypsum plaster.) Stucco plaster is a plaster mixed with a dilute solution of glue, and it is usually slow setting. An addition of alum or borax tends to increase the hardness of a plaster. Gypsum plasters are used for interior wall plasters, stucco work, architectural ornamentation, etc.

Gypsum blocks, tile, and plaster boards are made from gypsum wall plaster. These materials are used to some extent in building construction. They are light in weight, have good fire resisting qualities, are strong enough for many types of construction, and are easily sawn or cut to the desired shape. Gypsum wall plaster, mixed with water and fine cinders or wooden chips, has been used in making floors for buildings. These floors are usually lighter in weight but less strong and less fire resistant than concrete floors.

B. NATURAL CEMENT

5. Definition.—Natural cement is the finely pulverized product resulting from the calcination of a natural argillaceous limestone at a temperature below fusion. This temperature should be high enough (from 1,850 to 2,350 degrees Fahrenheit, to drive off the carbon dioxide as a gas, decompose the clay, and cause the formation of aluminates, ferrites, and silicates. The burned stone must be finely ground before it exhibits any hydraulic properties.

6. Manufacture of Natural Cement.—The rock used is a natural clayey limestone containing from 15 to 35 per cent of clayey material, 10 to 20 per cent of the clayey matter being silica, and the balance alumina and iron oxide. This rock should occur in large deposits and should be fairly uniform in composition.

The ordinary form of kiln in which the stone is burned is a vertical steel cylinder lined with firebrick and open at the top. It is about 30 or 40 ft. high and from 10 to 15 ft. in diameter.

Thick layers of limestone and thin layers of soft coal are alternately dumped into the top of the kiln and the burned clinker is removed through a door at the bottom. As the limestone descends in the kiln it first loses its water and then, when a temperature of about 1,400 degrees Fahrenheit is reached, the magnesium carbonate begins to decompose, freeing carbon dioxide. At a temperature of about 1,650 degrees Fahrenheit the carbon dioxide is driven off from the calcium carbonate. The clay decomposes at a slightly higher temperature and sets free alumina and iron oxide which combine with the lime and magnesia and, when the temperature is raised to about 2,200 degrees Fahrenheit, form silicates of lime and magnesia. The kilns are run continuously.

If the stones were all perfectly burned, the weight of the cement produced would be equal to the weight of the raw materials minus that of the carbon dioxide and the water in the rock. However, on account of the overburning and underburning of some of the rock, about 25 per cent of the clinker cannot be used for cement. The amount of soft coal required is about 30 lb. per barrel of cement made.

After the clinker is removed from the kiln, it is allowed to stand in air for a time so that any underburned clinker will be slaked before grinding. The slaking may be hastened by steaming in a closed vessel. After slaking, the clinker is crushed in a stone crusher and then ground to a fine powder in other forms of grinding machinery. Finally, the cement is packed in sacks or barrels for shipment.

7. Properties of Natural Cement.—The chemical composition of natural cement is about as follows: silica, SiO_2, 20 to 30 per per cent; lime, CaO, 30 to 60 per cent; magnesia, MgO, 1 to 25 per cent; alumina, Al_2O_3, 5 to 15 per cent; iron oxide, Fe_2O_3, 1 to 10 per cent; and small percentages of carbon dioxide, water, alkali, and sulphur trioxide. Because of differences in the chemical composition of the rocks used and in the degree of calcination, the chemical properties are very variable.

The specific gravity of natural cement varies usually from 2.70 to 3.10 with an average of about 2.95.

When mixed with water, natural cement will set either under water or in air. It usually sets more rapidly than portland cement. Allowing the cement to aërate for some time will cause it to set less rapidly and also to be less strong. Consequently, natural cement should not be stored exposed to air for more than a few weeks before using. The addition of gypsum or plaster of paris will retard the set somewhat. The standard specifications require initial set to occur in not less than 10 minutes and final set in not less than 30 minutes nor more than 3 hours, when the Vicat needle is used. The cement should be sound, and test pats stored in air and in water at normal temperature should remain firm and hard and show no signs of distortion, checking, cracking, or disintegrating.

The cement must be ground so fine that 90 per cent of it will pass a standard 100-mesh sieve and 70 per cent of it will pass a standard 200-mesh sieve. In general, the finer the cement is ground, the stronger it will be.

Natural cement pastes and mortars are about half as strong as corresponding portland cement pastes and mortars in tension, and only about a third as strong in compression. When tested in compression at the age of 1 month, neat natural cement cubes should average 800 lb. per square inch or more, and 1:2 sand mortar cubes should average more than 500 lb. per square inch. The A. S. T. M. standard specifications for natural cement require that the tensile strength of neat natural cement and 1:3 standard sand mortar should be equal to or more than the following values:

Tensile Strength

Age and storage	Neat cement, pounds per square inch	1:3 standard sand mortar, pounds per square inch
24 hours in moist air	75	
24 hours in moist air, 6 days in water	150	50
24 hours in moist air, 27 days in water	250	125

8. Uses of Natural Cement.—Natural cement is used sometimes in structural works where mass and weight, rather than strength, are required, as in sewers, conduits, massive foundations, pavement foundations, sidewalks, and rarely in large masonry dams, abutments, etc. Natural cement, when mixed with sand or with lime and sand, makes a suitable mortar for brick and stone masonry that is not subjected to heavy loads. Natural cement should not be used in exposed places or under water or where it will be exposed to the action of frost before the concrete has set and dried. At the present time the use of natural cement is decreasing, due to the decrease in cost and the increase in use of the better and stronger portland cement.

C. MISCELLANEOUS CEMENTS

9. Natural Puzzolan Cement.—Natural puzzolan cement is the finely pulverized product made by a mechanical mixture of fused argillaceous material and hydrated lime. All natural puzzolanic materials of commercial importance are taken from the deposits of volcanic ash. Hydrated lime must be added to this volcanic dust to form a hydraulic cement. As most deposits of puzzolana vary in quality and fitness for use, a careful selection must be made at the quarry to keep out objectionable materials. The selected material is ground very fine. It is usually mixed with hydrated lime (and also sand) at the place where it is to be used for structural purposes.

Good puzzolan cement mortar of a 1:3 mix is about as strong in compression as a like mortar made with portland cement, but it is only about 70 per cent as strong in tension.

Natural puzzolan cements were much used in the days of the Roman Empire and at that time they were the only known cementing materials. At the present time these cements are used but very little in construction work.

10. Slag Cement.—Slag cement is practically the same as puzzolan cement except that blast-furnace slag is used in place of the puzzolan rock. The slag must be a basic slag, such as is produced in the reduction of iron ores, and the slag should be cooled rapidly when it is taken from the blast furnace so that it will become broken up into small pieces which are easily handled by the grinding machinery. The granulation of the slag tends to make a stronger cement and also to reduce the amount of undesirable sulphides present. The slag is well dried, finely ground, and then thoroughly mixed with the proper proportion of hydrated lime.

The specific gravity of slag cement varies from 2.70 to 2.85 and it is about as finely ground as portland cement, though it is usually slower setting. The following table will give an idea of the strength of good slag cement:

STRENGTH OF SLAG CEMENT

Mix	Age	Tension, pounds per square inch	Compression, pounds per square inch
Neat paste	1 month	450	3,000
1:3 sand mortar	1 month	175	700

The uses of slag cement are usually limited to the unimportant parts of structural works which are not exposed and which do not require great strength. Slag cement is used but very little at the present time.

11. Magnesia or Sorel Cement.—This cement is magnesium oxide, MgO, which, when mixed with a proper solution of magnesium chloride, $MgCl_2$, forms oxychloride of magnesium. This is often called Sorel stone after M. Sorel, a Frenchman, who was the first to note that this cement exhibited very strong hydraulic properties. Much of the magnesium oxide is obtained from Greece, though some is produced in Canada and the United States.

Oxychloride of magnesium is probably the strongest and hardest artificial stone known at the present time. Magnesia cement forms a good, strong, tough mortar when mixed with sand, sawdust, asbestos, and other inert materials. The strength of neat pastes and mortars made with this cement is about twice that obtained by using portland cement.

Magnesia cement is used for floors in buildings and railway carriages, for stucco work, architectural ornamentation, etc. This cement should not be used in water or where it will be exposed to a great amount of moisture.

CHAPTER II

LIMES AND LIME MORTARS

A. LIMES

12. Definitions and Classifications. 1. *Quicklime.*—A white oxide of calcium, CaO, or a mixture of calcium and magnesium oxides, CaO and MgO.

(*a*) High-calcium lime contains 90 per cent or more of calcium oxide.

(*b*) Calcium lime contains from 85 to 90 per cent of calcium oxide.

(*c*) Magnesium lime contains from 10 to 25 per cent of magnesium oxide.

(*d*) Dolomitic lime contains more than 25 per cent of magnesium oxide.

Quicklime may be divided into two general grades as follows:

(*a*) Selected Lime.—A well-burned lime containing no ashes, clinker, or other foreign material. It contains 90 per cent or more of calcium and magnesium oxides and less than 3 per cent of carbon dioxide. Sometimes called "white" lime.

(*b*) Run-of-kiln Lime.—A well-burned lime containing 85 per cent or more of calcium and magnesium oxides and less than 5 per cent of carbon dioxide.

2. *Hydrated Lime.*—A quicklime to which just enough water has been added to produce a complete slaking.

3. *Hydraulic Lime.*—Obtained from the calcination of an ordinary limestone containing from 10 to 20 per cent of clay.

13. Manufacture of Quicklime.—The essentials of this process are the heating of a pure or magnesium limestone ($CaCO_3$ or $CaCO_3$ and $MgCO_3$) until the water in the stone is evaporated; then raising the temperature high enough for chemical dissociation and the subsequent driving off of the carbon dioxide as a gas; and leaving the oxides of calcium and magnesium. The maximum temperature required varies from 1,650 to 2,350 degrees Fahrenheit, depending upon the kind of limestone and the impurities present.

The fuels used in lime burning are wood, bituminous (soft) coal, and producer gas. The soft coal is not so good as wood because it burns with a much shorter flame, thus causing a more uneven heat distribution.

The kilns used in heating the limestone are usually vertical kilns of the intermittent or continuous types. Continuous kilns may be of the mixed feed, separate feed, or ring types.

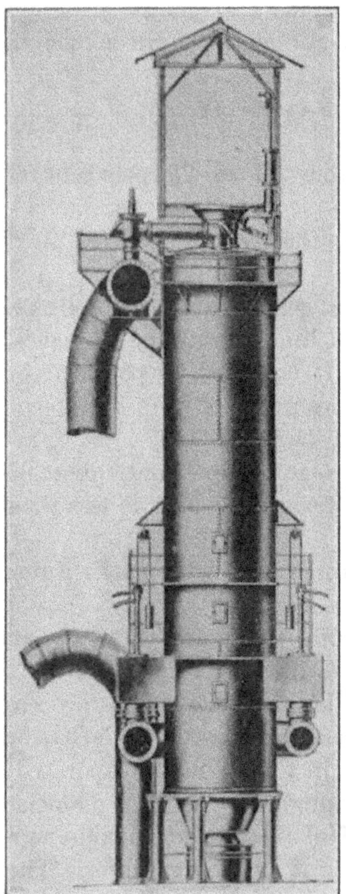

FIG. 2.—Continuous gas fired lime kiln (*Glamorgan Pipe & Foundry Co.*)

In a vertical kiln, the limestone is fed in at the top end and, as it descends, it first loses its water by evaporation; then the stone undergoes dissociation, the carbon dioxide passing off as a gas; and finally the calcined lime collects in the lower portion of the kiln, from which place it is withdrawn from time to time and the underburned and overburned materials sorted out. The cooled lime is sometimes ground to a powder (fine enough to pass an 80-mesh sieve) before being placed on the market.

In the intermittent or old-style form of kiln, the limestone is rarely ever uniformly burned and the fuel consumption is large. Consequently, these kilns are not used very much.

In the mixed-feed type of kiln, the mixture of bituminous coal and limestone is fed in at the top and the calcined material removed through a door at the bottom of the kiln. Often the limestone and fuel are charged in the kiln in alternate layers. The fuel consumption of this kind of kiln amounts to from 15 to 25 per cent of the weight of the lime produced.

The vertical kiln with the separate feed is made of steel and lined with firebrick, and it is so designed that the limestone does not come in contact with the fuel during the burning. The fuel is burned in a grate which is attached to the side of the kiln, and so arranged that the heat will ascend into the stack.

Compared with the separate feed kilns, the mixed feed kilns are

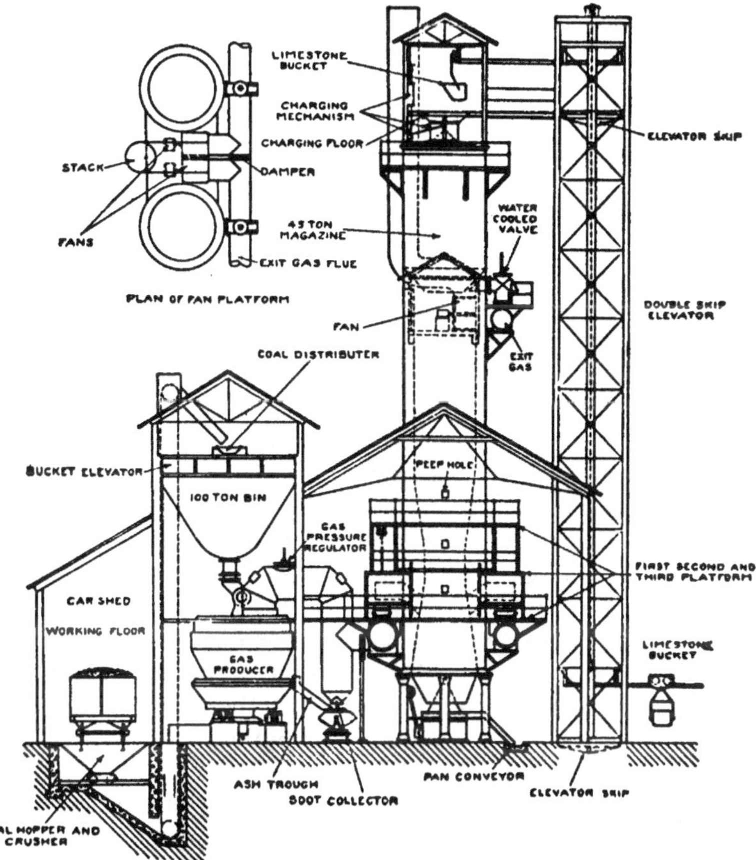

FIG. 3.—Mount continuous lime kiln plant. (*Glamorgan Pipe & Foundry Co.*)

cheaper to construct, a little more rapid in operation, and more economical in fuel, but they do not produce so high a quality of lime.

The ring or chamber type of kiln is much used in Germany. This kiln consists of a series of chambers grouped around a central stack, each chamber being connected with the stack and with the other adjacent chambers by a system of flues. The kiln is charged at the top with a mixture of fuel and limestone. In the burning, the flue system is so used that the hot gases generated in one chamber will pass through the other chambers before ascending the stack, thus causing a preheating of the limestone in the other chambers. This kind of kiln is economical in the amount of fuel used.

14. Manufacture of Hydrated Lime.—The commercial hydrated lime is made by crushing lump quicklime to lumps about half an inch in size or less (some factories crush the lime so fine that the larger portion will pass a 50-mesh sieve). The crushed lime is mixed with just enough water to secure a complete hydration. This mixing is usually done with machinery. The lumps of unhydrated lime and other impurities are removed by screening or by air separation, and the hydrated lime, in the form of a very fine powder, is packed in bags weighing about 100 lb. For every 56 parts of pure quicklime, 18 parts of water are required for the hydration. During the mixing, considerable heat is evolved and the lime increases in volume. The final product is a fine powder having about three times the volume of the original quicklime. The chemical formula is $Ca(OH)_2$.

15. Manufacture of Hydraulic Lime.—Hydraulic limes include all of those cementing materials (made by burning siliceous or argillaceous limestones) whose clinker after calcination contains so large a percentage of lime silicate (with or without lime aluminates or ferrites) as to give hydraulic properties to the product, but which at the same time contain normally so much free lime that the mass of clinker will slake on the addition of water.

The limestone rock used should be such that, after the silica has combined with the lime during calcination, enough free lime remains to disintegrate the kiln product by its own expansion when it is slaked. Such a limestone usually contains from 40 to 50 per cent of lime; about 1 per cent of magnesia; from 7 to 17 per cent of silica; and about 1 per cent of alumina and iron oxide.

The hydraulic limes are manufactured in continuous kilns in the same way as quicklime except that a higher temperature (never less than 1,850 degrees Fahrenheit) is required. After the burning, the lumps of lime are removed from the kiln and slaked in the same way as quicklime, great care being taken to use just the right amount of water and no more, as an excess of water would cause the lime to harden. The expansion of the quicklime in slaking breaks up the lumps into a fine powder which consists principally of lime silicate with about 25 to 33 per cent of hydrated lime. The lime is then screened through a 50-mesh sieve and placed in bags.

The underburned limestone and overburned materials (known as grappiers), which are left after the hydraulic lime is slaked and screened, are ground to a fine powder and sold as "grappier" cement. This cement is of value according to the proportion of lime silicate contained in it.

Lafarge cement is a hydraulic grappier cement made at Tiel, France.

16. Properties of Quicklime.—Before using, quicklime must be slaked by the addition of water which causes the calcium oxide to change to calcium hydroxide, $Ca(OH_2)$. As great heat is evolved when quicklime is slaked, it should be stored so that the heat caused by an accidental slaking of a part of the lime will not cause a fire. The rate of hydration and the evolution of heat vary according to the purity of the lime and the percentage of calcium oxide present. The high-calcium quicklimes slake more rapidly and generate more heat than the other quicklimes.

When the slaked lime is exposed to the air, it gradually absorbs carbon dioxide and changes from calcium hydroxide to calcium carbonate (limestone) and water. Dry carbon dioxide will not react with dry hydrated lime, hence an excess of water (moisture) must be present.

On account of the large shrinkage in the hardening of lime paste, sand or some other inert material must be added to reduce the shrinkage and cracking. The proportions are usually 1 part of lime to from 2 to 4 parts of sand. The plasticity (quality of being spread easily and smoothly with a mason's tool) or sand-carrying capacity of a lime may be expressed by the number of parts of sand which can be mixed with 1 part of lime paste without making the mortar too stiff or "short" to work well with a trowel. The sand-carrying capacity of a lime appears to vary with its purity and calcium content, the high-calcium limes being able to carry the most sand. The yield of a lime is the volume of paste of a given consistency produced by a unit weight of dry quicklime. The greater the purity and the higher the calcium content of the lime, the greater the yield. The hardness seems to vary inversely, and the shrinkage to vary directly, with the purity of the lime and the percentage of calcium oxide present.

The A. S. T. M. specifications for quicklime in regard to

physical properties and tests are: "An average 5-lb. sample shall be put into a box and slaked by an experienced operator with sufficient water to produce the maximum quantity of lime putty, care being taken to avoid "burning" or "drowning" the lime. It shall be allowed to stand for 24 hours and then washed through a 20-mesh sieve by a stream of water having a moderate pressure. No material shall be rubbed through the screens. Not over 3 per cent of the weight of the selected quicklime nor over 5 per cent of the weight of the run-of-kiln quicklime shall be retained on the sieve. The sample of lump lime taken for this test shall be broken to pass a 1-in. screen and be retained on a ¼-in. screen. Pulverized lime should be tested as received."

17. Properties of Hydrated Lime.—Hydrated lime is the same as ordinary quicklime which has been properly slaked and, therefore, it should have the same physical properties. However, it has been observed that hydrated lime makes a mortar that is stronger, more rapid setting, and which shrinks less than the ordinary quicklime mortars. The sand-carrying capacity and yield of hydrated lime are usually less than that of quicklime. The better qualities of the hydrated lime may be due to the more perfect hydration.

The A. S. T. M. specifications for hydrated lime in regard to physical properties and tests are: "A 100-gram sample shall leave by weight a residue of not over 5 per cent on a standard 100-mesh sieve and not over 0.5 per cent on a standard 30-mesh sieve." "Hydrated lime shall be tested to determine its constancy of volume in the following manner: Equal parts of hydrated lime under test and volume-constant portland cement shall be thoroughly mixed together and gaged with water to form a paste. Only sufficient water shall be used to make the mixture workable. From this paste a pat about 3 in. in diameter and ½ in. thick at the center, tapering to a thin edge, shall be made on a clean glass plate about 4 in. square. This pat shall be allowed to harden 24 hours in moist air and shall be without popping, checking, cracking, warping, or disintegration after 5 hours' exposure to steam above boiling water in a loosely closed vessel."

18. Properties of Hydraulic Lime.—Hydraulic lime pastes and mortars are about as strong as those of natural cement. Compared with the values obtained from tests on portland cement pastes and mortars, the strength of hydraulic lime pastes and mortars is about 1/3 as strong in tension and about 1/4 as strong in compression. A 1:3 hydraulic lime and sand mortar is about 70 per cent as strong as the neat mix. The rate of gain in strength is very slow and the maximum strength is not reached in less than a year. Hydraulic limes are about five times as strong in compression as they are in tension. The above remarks are not true for the feebly hydraulic cements as those cements are very much weaker.

19. Uses of **Limes.**—About half of the lime made is used for various structural purposes, the remainder being used for other industrial purposes and arts. Most of the lime used for structural purposes is mixed with sand to form mortars for laying brick and stone masonry. A large amount of lime is used in plastering the walls and ceilings of buildings. Ordinary wall plaster is a mixture of lime and sand to which hair, fiber, etc. have been added. Some lime is used for whitewashing. A little lime is sometimes used in cement mortars to make them more plastic and impermeable.

Hydrated lime is used for the same structural purposes as quicklime, and it is more easily handled, stored, and shipped as there is no danger of air slaking. Hydrated lime requires no slaking before it is ready to be mixed with sand and water to form a mortar. It is sometimes used as an ingredient of portland cement mortars and concretes.

Hydraulic limes and grappier cements are sometimes used for the purposes of interior decoration. At one time they were much used in construction work, but they were replaced some time ago by the natural cements, and later by portland cement. Hydraulic limes are not suitable for use in underwater work and they are too slow setting for practical construction work.

B. LIME MORTARS

20. Lime **Mortar; Definition and** Materials.—Lime mortar is a mixture of slaked (hydrated) lime, usually in the form of a thick paste, sand, or other fine aggregate, and water.

The lime used is usually a quicklime which must be properly slaked or hydrated before the sand or other fine aggregate is added. In general, a high-calcium lime makes the strongest and best-working mortar for ordinary uses. Sometimes a hydrated lime (a lime which has been slaked by the manufacturer) in the form of a fine powder is used. This hydrated lime requires no slaking or other preparation and is ready to be mixed at once with the sand and water to form a mortar.

The water used should be clean and contain no materials, such as oils, acids, strong alkalis, vegetable matter, etc., which may be injurious to the mortar.

The sand used for lime mortar should be clean and sharp and be composed of rather small grains in preference to large ones. The sand should be free from all dirt, loam, clay, and vegetable matter as these impurities tend to decrease the strength and soundness of the mortar.

21. Slaking the Quicklime.—When quicklime is used, it must first be properly slaked before being mixed with the fine aggregate. It is important to secure a complete slaking of the lime and no more, because, if too much water is added, some of the binding power of the lime will be destroyed, and if too little water is used or proper care is not exercised by the workman, some of the lime may not be slaked. This unslaked lime may slake after the mortar is in place and cause bad results, especially if the mortar is used as a wall plaster. If the quicklime is properly slaked, the lime paste formed should have about three times the volume of the original quicklime. There are three general methods of slaking quicklime, namely, drowning, sprinkling, and air-slaking.

Slaking by the drowning method is the most common way. The lumps of quicklime are placed in a layer 6 or 8 in. deep in a water-tight box and then water is poured on the lumps. The water should be equal to about two and a half or three times the volume of the quicklime. If the proper amount of water is used, the lime will form a thick paste. With a high-calcium (quick-slaking) lime, it is better to add the water all at once, but with a magnesium (slow-slaking) lime, the water should be added gradually. As lime slakes best when hot, care should be taken not to chill the lime after it has begun to slake. Stirring may be necessary to break up some of the lumps, but care should be taken not to chill the lime and retard the slaking. "Burning" occurs when only a little water is present and this water is changed into steam by the heat produced. "Burning" tends to prevent a complete slaking of the lime.

Another method of slaking by drowning is to fill a water-tight box with about 8 in. of water and then add lumps of lime in sufficient quantity to form a thick paste. The mass must be stirred to assist in breaking up the lumps of lime.

Slaking by sprinkling consists of sprinkling a heap of quicklime with water equal to about one-third or one-fourth of the volume of the lime and then covering the mass with sand and allowing it to stand for a day or so. If the slaking is properly done, the hydrated lime will be in the form of a powder. This method requires extra care and expert labor and is, consequently, expensive.

Air slaking consists of spreading the quicklime in a thin layer and allowing it to slake by absorbing moisture from the air. Frequent stirring is required. This method produces a good quality of slaked lime, but is rarely used due to the large storage area, labor, and time required.

22. Proportioning and Mixing of Lime Mortar.—Sand should be added to the lime paste for four reasons:

1. To prevent excessive cracking and shrinking of the lime mortar when the water evaporates.
2. To give greater strength to the mortar.
3. To divide the lime paste into thin films and to make the mortar more porous, thus aiding in the absorption of carbon dioxide from the air which causes the lime to set or harden.
4. To reduce the cost.

The usual proportions vary from 2 to 4 parts of sand to 1 part of lime paste. With most sands and limes, the correct proportions will be from 2½ to 3 parts of sand to 1 part of lime paste by volume. Care should be taken to secure the proper proportions. The volume of the lime paste should be just a little more than enough to coat completely all of the sand grains and fill the voids.

In mixing the mortar, the lime paste is first spread out in a thin layer a few inches thick and the sand spread uniformly over the top. The lime paste and sand are then mixed by hoe or shovel until the mass is of a uniform color. A little water should be added, if necessary, to make the mortar of the proper consistency. Thorough mixing is required to make a good mortar.

When hydrated lime in the form of a powder is used, the lime and sand should be mixed dry until of a uniform color and then sufficient water should be added and the whole mixed until the mortar is of the proper consistency.

If too much sand has been used the mortar will be "short"

and "stiff" and will not work properly; while if too much lime paste is used, the mortar will be too sticky to work properly. A mason can tell very quickly whether the mortar is correctly proportioned or not when he starts to use the mortar in his work. The proportions which give the best working mortar are also the best proportions in regard to strength, hardening, and other properties (except when clay or loam is used instead of sand).

About 210 lb. of good quicklime are required to make a cubic yard of 1:3 lime mortar.

23. Properties of Lime Mortar.—Lime mortar has the important property of "setting" or "hardening" when the water evaporates and the lime absorbs carbon dioxide from the air thus forming calcium carbonate. This setting takes place very slowly, especially if the mortar is placed in thick layers or in places where it is difficult for the air to reach, and sometimes many years are required for the hydrated lime to change to calcium carbonate.

In a lime mortar, an excess of lime paste delays the hardening, increases the shrinkage, decreases the compressive strength, and makes the mortar sticky. An excess of sand makes the mortar "short" and hard to work with a mason's tools besides decreasing the strength of the mortar.

The freezing of lime mortar delays the evaporation of the water and thus delays the absorption of carbon dioxide from the air. The expansion of the water due to the freezing may damage the mortar. Alternate freezing and thawing decrease the adhesive and cohesive strength.

A fine, sharp, clean sand gives the best results in a lime mortar.

Clay, loam, dirt, etc. decrease the strength of lime mortar, hence these materials should not be used.

Oils, acids, strong alkalis, vegetable matter, etc. decrease the strength and hardening qualities of a lime mortar.

The tensile strength of a good 1:3 lime mortar, 1 month old, varies from 30 to 60 lb. per square inch. When it is 6 months old, the strength will probably be from 10 to 15 lb. more per square inch.

The compressive strength of a good 1:3 lime mortar, at the age of 1 month, will probably be between 150 and 400 lb. per square inch., while at the age of 6 months the strength may vary from 170 to 750 lb. per square inch. The strength of a lime mortar depends upon the quality of the lime and the sand and upon the care taken during the mixing, molding, storing, and testing.

A magnesium lime mortar is usually stronger and quicker setting than a high-calcium lime mortar.

24. Common Lime or Wall Plaster.—Common lime or wall plaster is a lime sand mortar in which hair, fiber, or some similar material has been thoroughly mixed. The hair or fiber is added to keep the plaster from shrinking and cracking when it sets and hardens on the wall.

Wall plaster is usually applied in two coats. The first or rough coat is put on about half an inch thick. It consists of about a 1:3 lime mortar to which the fiber, etc. have been added. The exposed surface is troweled smooth, but no effort is taken to make a very smooth surface.

The second or finishing coat is added after the first coat has dried. This finishing coat consists of a rich mortar (a 1:1 or 1:2 mix) made of a very white lime paste and a fine, sharp, clean, light-colored sand. This coat is applied in a very thin layer and care is taken to secure a very smooth-finished surface.

It is important that the quicklime used in a wall plaster shall be thoroughly slaked before it is placed on the wall. This is usually made sure of by allowing the plaster to remain in a water-tight box for several days before it is applied to the wall. If any unslaked lime is placed in the wall, it will absorb moisture, slake, expand, and form "blisters" on the wall surface, often injuring the wall so much that the plastering has to be done over again.

A "whitewash" is a thin paste made of white quicklime and water which is applied to the wall or other surface by means of a brush. As many coats as desired may be applied. A "whitewash" serves the same purpose as a cheap paint.

25. Uses of Lime Mortar.—Lime mortar is used as a mortar for stone and brick masonry, where the mortar can be placed in comparatively thin layers and the walls are not very thick, and where great strength is not required. Lime mortar should not be used in massive masonry, under water, or in a wet soil, as the lime will not harden unless it can absorb carbon dioxide from the air.

In places where great strength is required, a portland cement mortar should be used.

Lime mortar is sometimes mixed with portland cement mortar to make the portland cement mortar easier to work and also where a mortar stronger than lime mortar is required.

Lime mortar is much used as a wall plaster and for stucco work, etc.

CHAPTER III

PORTLAND CEMENT

A. DEFINITION AND CLASSIFICATION

26. Definition of Portland Cement.—Portland cement is the product obtained by finely pulverizing clinker produced by calcining to incipient fusion, an intimate and properly proportioned mixture of argillaceous and calcareous materials, with no additions subsequent to calcination excepting water and calcined or uncalcined gypsum (Am. Soc. Test. Mat.).

27. Classification of the Principal Cementing Materials.—At the present time the knowledge of cement chemistry is not complete enough to allow of the classification of cementing materials according to their chemical properties. However, the different kinds of cementing materials may be classified according to their methods of manufacture and physical properties. The following classification brings out the main differences in the manufacturing methods and the slaking and hydraulic properties of the five main-cementing materials.

1. *Common Lime.*—Made by burning relatively pure limestone at a very low temperature. It will slake when mixed with water and it has no hydraulic properties.
2. *Hydraulic Lime.*—Made by burning slightly argillaceous limestone at a low temperature. It will slake slowly and has feebly hydraulic properties.
3. *Natural Cement.*—Made by burning argillaceous limestone at a comparatively high temperature. It will not slake but it has hydraulic properties when ground.
4. *Portland Cement.*—Made by burning an artificial mixture of argillaceous and calcareous materials to a temperature of incipient fusion. It will not slake but it has very marked hydraulic properties when finely ground.
5. *Puzzolan or Slag Cement.*—Made by mixing slaked lime with granulated blast-furnace slag or a natural puzzolanic material. It will not slake but possesses hydraulic properties when ground.

B. MANUFACTURE OF PORTLAND CEMENT

28. Raw Materials.—A large number of materials are available for use in the manufacture of Portland cement. The following are the materials most commonly used and they are arranged about in the order of their importance.

Argillaceous Materials	Calcareous Materials
Argillaceous limestone (cement rock)	and pure limestone
Clay or shale	and pure limestone
Clay or shale	and marl
Blast-furnace slag	and pure limestone
Clay or shale	and chalk or chalky limestone
Clay or shale	and alkali waste

Cement rock is a soft, impure, argillaceous limestone containing about 20 per cent of clay and 70 per cent of calcium carbonate.

Limestone suitable for cement manufacture consists principally of calcium carbonate (90 per cent or more) with small percentages of silica, aluminium and iron oxides, magnesium carbonate, sulphur, and various alkalis.

Marl is a deposit of soft and comparatively pure limestone usually found in the beds of extinct and existing lakes.

Shales are a soft rock composed chiefly of silica, alumina, and iron oxide.

Clays result from decayed shales and, consequently, have about the same chemical composition with a little more water.

Slate is a form of shale.

Blast-furnace slag is a fusible silicate formed during the reduction of the iron ore in a blast furnace by the combination of the fluxing material (limestone, etc.) with the earthy matter (gangue) of the ore.

Chalk is a soft variety of calcium carbonate formed from the remains of minute organisms. It also contains small percentages of silica, alumina, and magnesia.

Alkali waste is the precipitated calcium carbonate obtained during the manufacture of caustic soda by the Leblanc process.

In order to secure the proper chemical combinations in the kiln, all of the calcareous materials should be free from quartz or sand and should contain but little sulphur or magnesium carbonate. The clays should also be free from sand and harmful impurities. Soft limestones are more easily ground than hard ones.

29. Proportioning of the Raw Materials.—Portland cement has a complex chemical composition consisting for the most part of tricalcium silicate ($3CaOSiO_2$), dicalcium silicate ($2CaOSiO_2$), and tricalcium aluminate ($3CaOAl_2O_3$), with small amounts of other compounds, resulting from the burning of the calcium carbonates, silicates, and alumina. Hence, the proportions of argillaceous and calcareous materials must be carefully chosen in order to secure the proper results. Eckel's formula for the correct theoretical proportions for Portland cement is:

$$\frac{2.8\,\text{Silica}\,(SiO_2) + 1.1\,\text{Alumina}\,(Al_2O_3) + 0.7\,\text{Iron Oxide}\,(Fe_2O_3)}{1.0\,\text{Lime}\,(CaO) + 1.4\,\text{Magnesia}\,(MgO)} = 1.$$

An average portland cement contains about 22.0 per cent of

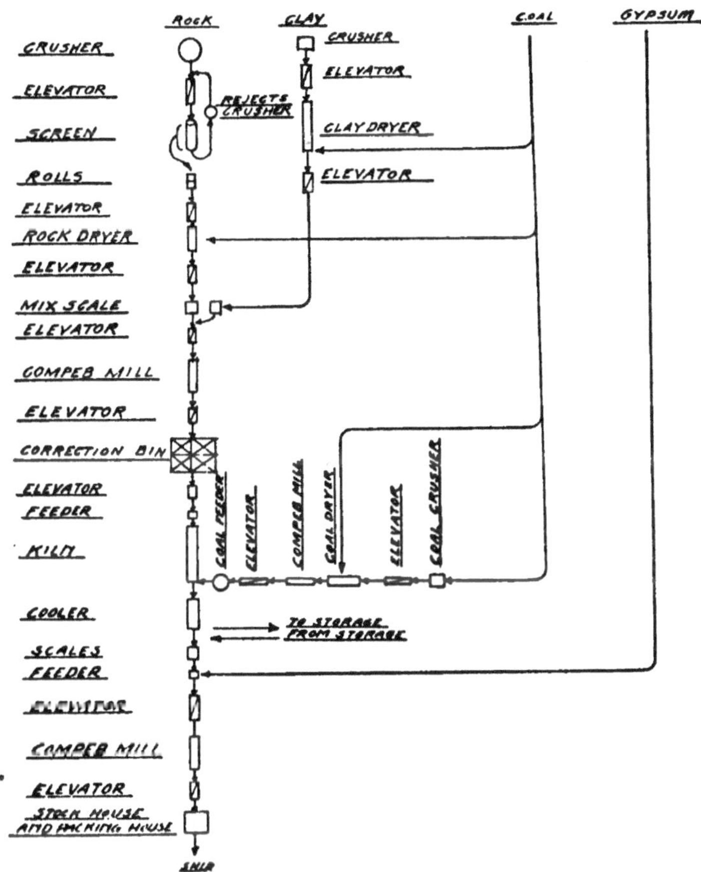

FIG. 4.—Flow sheet for a dry process Portland cement plant. (*Allis-Chalmers Mfg. Co.*)

silica, 7.4 per cent alumina, 3.0 per cent iron oxide, 62.0 per cent lime, 1.75 per cent magnesium oxide, 1.3 per cent of sulphuric acid, and about 1.0 per cent of alkalis.

An excess of lime makes the cement unsound, while too little lime causes the cement to be quick setting and weak. In order to avoid an excess, the lime content is kept a little below the value given by the formula.

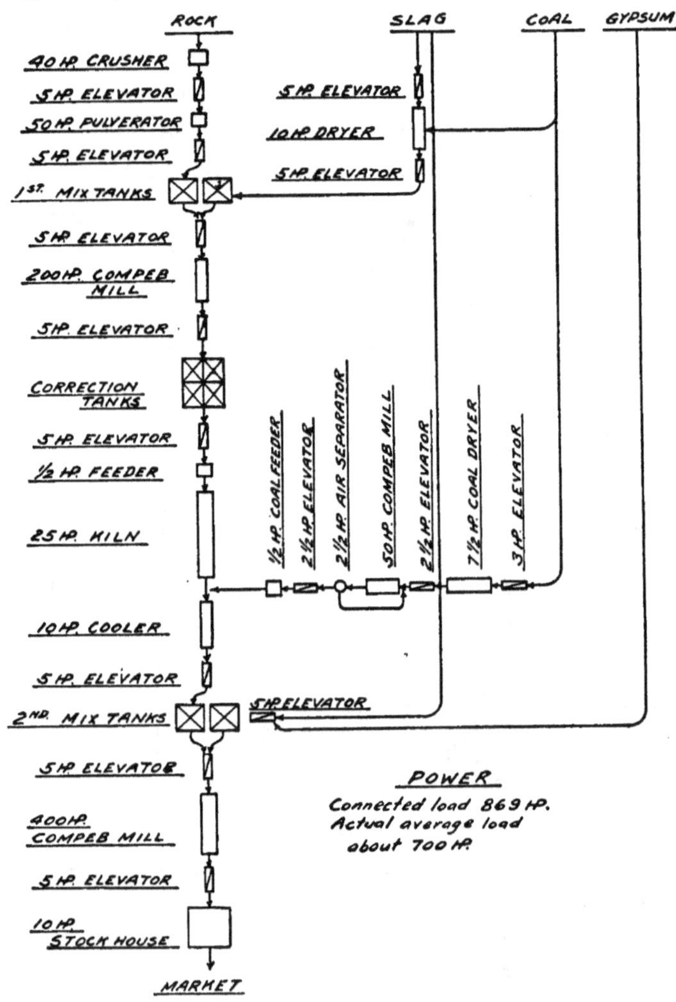

Fig. 5.—Flow sheet for a thousand barrel dry process slag Portland cement plant. (*Allis-Chalmers Mfg. Co.*)

The amount of alumina present has a large effect on the clinkering temperature, the more the alumina the lower the temperature required. Large amounts of alumina tend to make the cement quick setting and weak as well as to render the cement more liable to disintegration when exposed to the action of sea water.

Magnesia, up to 4 or 5 per cent, appears to have no bad effect on the cement, but larger amounts are thought to be injurious.

Calcium sulphate, as gypsum or plaster of Paris, is added to the cement after burning to retard the set. The amount is usually less than three per cent.

Sulphuric acid (SO_3) is limited to 200 per cent by the specifications.

30. Outline of the Dry Process of Manufacture.—The dry process of manufacture of Portland cement, which is the most important process of manufacture, consists of the following steps (though not always in the exact order as given):

1. Securing the raw materials.
2. Crushing the raw materials.
3. Drying the raw materials.
4. Grinding the raw materials.[1]
5. Proportioning and mixing.
6. Finely grinding the materials.[1]
7. Burning the materials.
8. Cooling the clinker.
9. Addition of the retarder.
10. Grinding clinker to a very fine powder.
11. Seasoning the cement.
12. Packing the cement for shipment.

31. The First Six Steps of the Dry Process of Manufacture.—1. Most of the raw materials are obtained by quarrying, after which they are transported to the factory. Blasting is required for the harder materials, while the softer materials may be excavated with a steam shovel. Marl is often dredged. Blast-furnace slag is secured from the blast furnace.

2. Nearly all of the raw materials need to be crushed to smaller sizes before they can be handled by the grinders. The crushing is done by means of jaw, roller, or gyratory crushers and often both large and small crushers are used to make the materials fine enough for the grinders. The gyratory type of crusher is used more than the other types because it is more economical in operation. The raw materials are usually crushed so that they will pass through a 2-in. ring.

3. The drying is usually done in a revolving, hollow, steel cylinder about 5 ft. in diameter and 50 ft. long and which is inclined a little to the horizontal. The materials enter at the upper end and pass out of the lower end. The dryer removes the water from the materials.

[1] NOTE.—In the newer types of cement mills, both the coarse and fine grinding (processes 4 and 6) are done in one mill.

4. The preliminary grinding is often done in a ball mill which is a short, closed, hollow cylinder that revolves about its longitud-

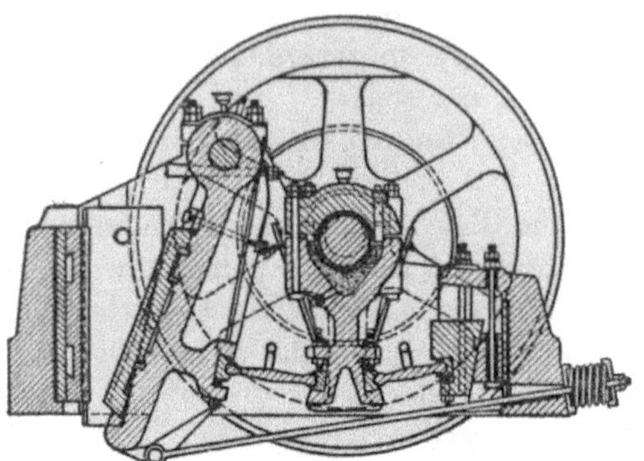

Fig. 6.—Jaw crusher. (*Allis-Chalmers Mfg. Co.*)

Fig. 7.—Gyratory crusher. (*Allis-Chalmers Mfg. Co.*)

inal axis. The grinding is done by a number of hard steel balls, from 3 to 5 in. in diameter, placed in the cylinder. The materials are ground so that they will pass a 20-mesh sieve.[1]

[1] Note.—In the later type of cement plants, both the coarse and fine grinding are done in one mill, called a compeb mill. This mill is divided into two parts. The coarse grinding is done in the first part, after which the material passes into the second part for the fine grinding. The use of this type of mill eliminates process 4 (grinding the raw materials).

Fig. 8.—Direct heat rotary dryer. (*Allis-Chalmers Mfg. Co.*)

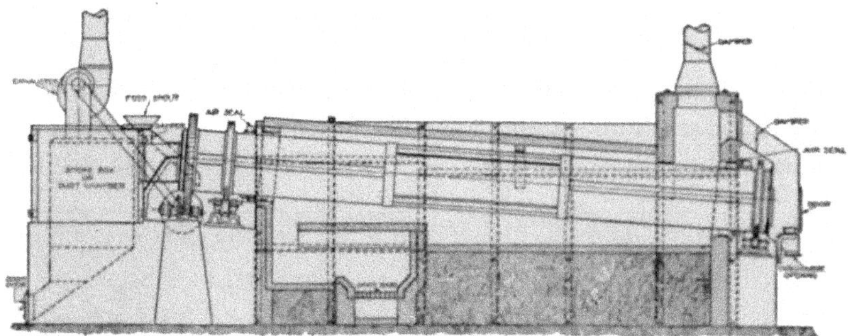

Fig. 9.—*I*ndirect direct heat rotary dryer. (*Allis-Chalmers Mfg. Co.*)

Fig. 10.—Compeb mill. (*Allis-Chalmers Mfg. Co.*)

5. The materials then pass through a weighing machine where they are weighed out in the correct proportions and then dumped in a mixing hopper where they are thoroughly mixed together.

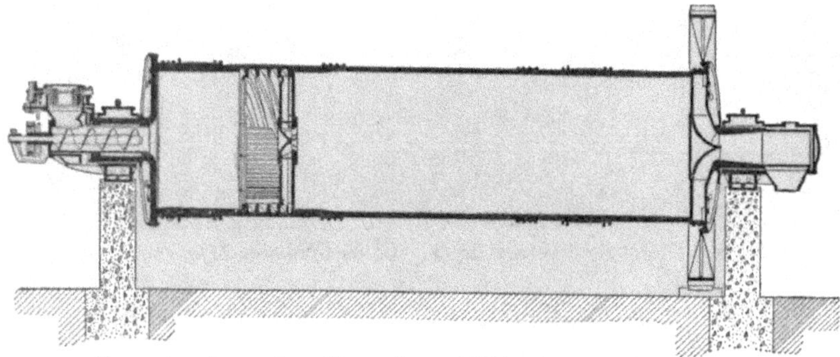

FIG. 11.—Compeb mill section. (*Allis-Chalmers Mfg. Co.*)

6. From the mixing hopper, the materials pass into a mill (such as a tube, Gates, Fuller-Lehigh, or Griffin mill, etc.) for a finer grinding. The tube mill, which is more often used than any

FIG. 12.—Gates tube mill. (*Allis-Chalmers Mfg. Co.*)

other, is a closed steel cylinder about 5 ft. in diameter and 22 ft. long which is lined with a material (that has a high abrasive resistance) such as chilled cast iron or trap rock. The grinding is done by a number of flint rocks, about the size of goose eggs, which fill the mill about half full. The materials are ground so fine that about 95 per cent of the powder will pass a 100-mesh sieve. (See footnote, page 25.)

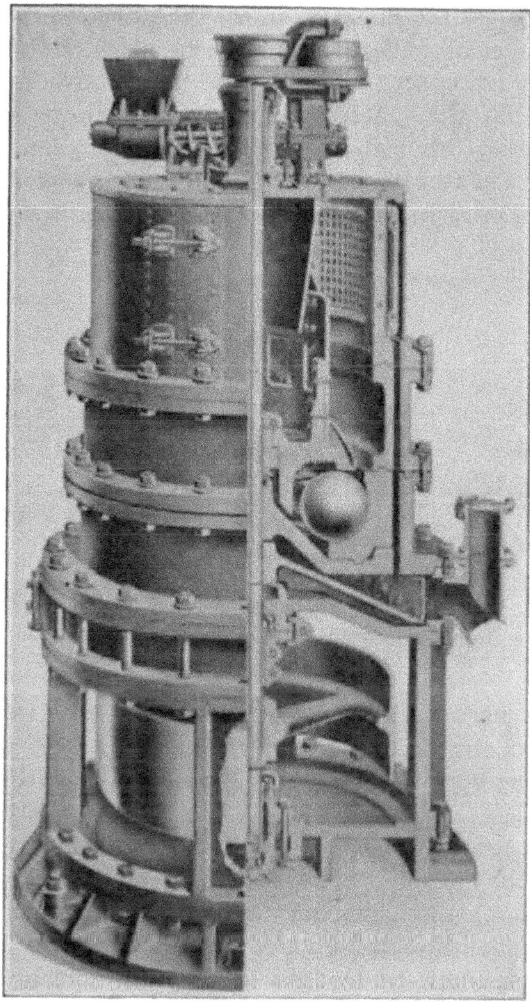

FIG. 13.—Fuller-Lehigh mill. (*Fuller-Lehigh Co.*)

32. The Remaining Six Steps of the Dry Process of Manufacture.—7. The finely ground material from the tube mills passes into the upper end of a rotary type of cement kiln where it is burned. The rotary kiln is a long steel cylinder, 6 to 9 ft. in diameter and from 100 to 150 ft. in length, which is slightly inclined to the horizontal and is so made that it can be slowly rotated. The fuel used is a finely powdered coal which is blown through a nozzle inserted in the lower end of the kiln. A brick flue, leading to a smokestack, is attached to the upper end of the kiln. Soon after the material enters the upper end of the kiln it balls up in small balls and, as it moves slowly down the kiln, the water is evaporated and the most of the carbon dioxide is driven off. As the material approaches the lower end of the kiln, all of the carbon dioxide, sulphur, and organic matter is expelled. A few feet from the lower end the temperature reaches 2,900 to 3,100 degrees Fahrenheit and the little brown balls are

Fig. 14.—Rotary kiln with taper end. (*Allis-Chalmers Mfg. Co.*)

fused into a hard dark-colored clinker. The time required for the passage of the material through the kiln is four or five hours.

8. After the burning, the clinker is removed from the kiln and sprayed with a stream of water. Then the clinker is passed through a cooler and placed in the clinker storage bins.

9. When the clinker is removed from the clinker storage bins, it passes through a weighing machine where the retarder is added. The reason that something is added to retard the set of the cement is that the high-lime content would make the cement too quick setting for commercial use. The quantity added is about 2 per cent, usually, and never more than 3 per cent. Gypsum is usually used as a retarder though plaster of Paris is sometimes used.

10. After the retarder is added the clinker is ground to a very fine powder in a mill similar to the one used for finely grinding the material before burning. The clinker must be ground so fine that 78 per cent or more will pass a standard 200-mesh sieve.

11. After the final grinding, the cement is conveyed to a storage bin and allowed to season for a few weeks before being packed for shipment. These storage bins usually have a capacity varying from 1,000 to 5,000 bbl. each. The seasoning seems to improve the quality of the cement.

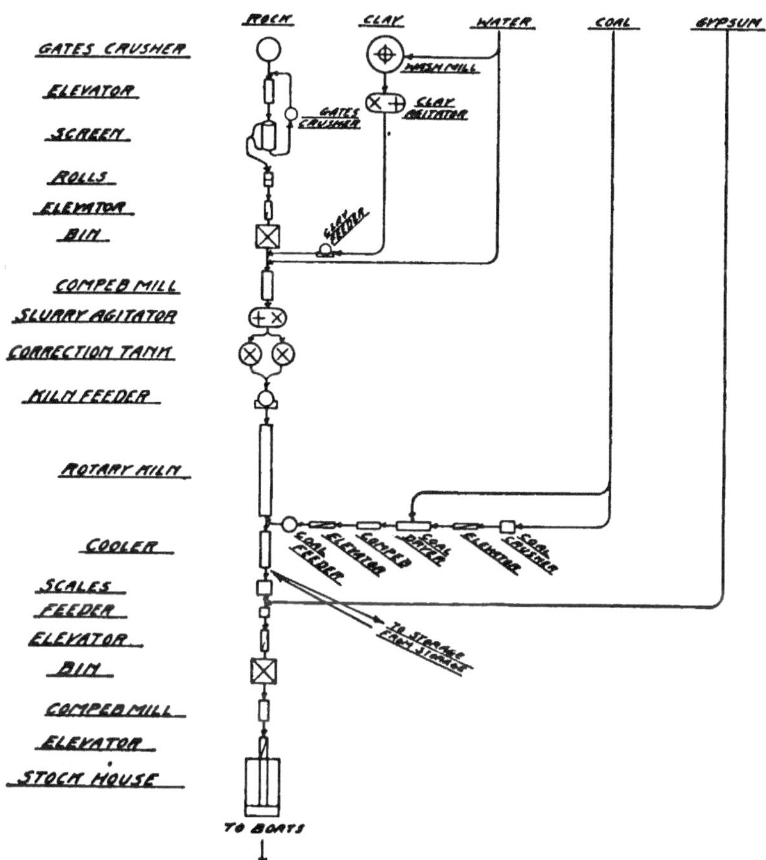

FIG. 15.—Flow sheet for a wet process Portland cement plant. (*Allis-Chalmers Mfg. Co.*)

12. For shipment, the cement is packed in bags or sacks holding about 94 lb. of cement, or in barrels which hold the equivalent of four sacks or 376 lb. of cement. Sometimes the cement is placed in bulk in a railroad car and so shipped.

33. The Wet Process of Manufacture.—The raw materials most commonly used in this process are clay and chalk or marl.

The clay is dried and then ground in an edge runner mill, while the other materials are ground in a wash mill where enough water is used to make them into a thin mud or slurry. Then the proper quantities of the materials are weighed out and mixed in a pug mill, the slurry pumped into a large vat, a chemical analysis made, and more materials added if necessary. The slurry is then pumped from the vat to a special rotary kiln in which it is burned. After the clinker is removed from this kiln, the process of manufacture is the same as that of the dry process.

The wet process of manufacture allows of better chemical control and easier grinding, but it requires more fuel for the burning. Because of this the dry process is usually cheaper and, cousequently, more often used.

C. PROPERTIES AND USES OF PORTLAND CEMENT

34. General.—Cement is valuable as a structural material because it has mechanical strength after hardening. In order to compare the mechanical strengths of different cements and their fitness for structural work, it is necessary to make standardized tests and laboratory experiments on some of the physical and mechanical properties. These qualities in the order of their importance are: soundness, strength, time of set, fineness, and specific gravity. The determinations of some of the chemical properties by standardized chemical tests aid in deciding whether a cement is suitable or not for structural purposes.

35. Chemical Constitution and Specifications.—The latest studies on the chemical constitution of Portland cement seem to indicate that portland cement is made up largely of three compounds, namely: tricalcium silicate ($3CaOSiO_2$), dicalcium silicate ($2CaOSiO_2$), and tricalcium aluminate ($3CaOAl_2O_3$). There are also small amounts of iron oxide (Fe_2O_3), magnesia (MgO), sulphur in the form of SO_3, alkalis, etc., together with a little lime (CaO) if the clinker is underburned. A perfectly burned cement clinker consists of about 36 per cent of tricalcium silicate, 33 per cent of dicalcium silicate, 21 per cent of tricalcium aluminate, and about 10 per cent of other compounds. If the cement clinker is not perfectly burned, there is less tricalcium silicate and more dicalcium silicate and usually some free lime.

The standard specifications for Portland cement specify that, in regard to chemical properties, the following limits shall not be exceeded:

Loss on ignition	4.00 per cent
Insoluble residue	0.85 per cent
Sulphuric anhydride (SO_3)	2.00 per cent
Magnesia (MgO)	5.00 per cent

36. Soundness.—Soundness is a necessary quality for cement that is to be used for structural purposes, as it is not desirable to use a cement that will later disintegrate and cause a failure of the structure. Unsoundness is usually shown by expansion after the cement has set, followed by disintegration. Free lime is the chief cause of unsoundness, causing the cement to expand and disintegrate. An excess of magnesia (more than 5 per cent) is thought to cause unsoundness. An excess of sulphate is thought to have a similar action in some cases. Seasoning helps in making cement sound by giving time for the complete hydration or carbonating of any free lime that is present. Unsoundness is shown by the cracking and disintegration of the cement after setting, due to the expansion of some of its constituents. The amount of sulphates added for a retarder should never be more than 3 per cent. Soundness is promoted by thorough seasoning, fine grinding, and by keeping the amounts of magnesia and sulphates low.

The specification requires that a pat of neat cement shall be kept in moist air for 24 hours and then exposed for 5 hours in an atmosphere of steam at a temperature between 98 and 100 degrees Centigrade on a suitable support 1 in. above the boiling water. This pat of neat cement should be about 3 in. in diameter, ½ in. thick at the center, and tapering to a thin edge. To pass the soundness test successfully, the pat should remain firm and hard and should show no signs of distortion, checking, cracking, or disintegrating. This test is commonly known as the "accelerated" test.

The old specifications required three pats to be tested. One was to be subjected to the accelerated test. One pat was to be kept in moist air for 24 hours and then in water for 27 days, and observed at intervals. The third pat was to be kept in moist air for 24 hours and then in air for 27 days, and observed at intervals. The temperature of the air and water was to be kept as near 70 degrees Fahrenheit as practicable.

37. Strength.—The tensile strength of cement has but little value as a measure of the suitability of the cement for structural purposes, but it is of value as a means of comparing different cements and also because of the relation of the tensile to the compressive strength. While the ratio of tensile strength to compressive strength varies for different cements and for different ages, it is generally true that a cement that is strong in tension is also strong in compression. Because of fewer difficulties in the making and testing of specimens and because of the lower cost and weight of testing machines required, tension tests have been standardized in preference to compression tests on cement.

A slight increase in the lime content increases the tensile strength a little. Fine grinding increases the strength of cement mortar but not that of the neat cement. An addition of more than 5 or 10 per cent of clay is injurious. Hydrated lime will decrease the strength of neat cement. A high-tensile strength does not indicate that the cement is sound. A large retrogression in the strength of cement is a bad sign.

The tensile test requirements for neat cement have been discontinued in the new specifications. The following are the old specifications (minimum requirements) for the strength of neat cement:

Storage	Strength, Pounds per Square Inch
1 day in moist air	175
1 day in moist air and 6 days in water	500
1 day in moist air and 27 days in water	600

The compressive strength of cement is the best criterion to use in choosing a cement for structural purposes, but for several reasons this test has not been standardized. The compressive strength of a good neat cement is about 10 times its tensile strength. The modulus of elasticity for cement in compression is not a constant because the stress strain curve is not a straight line for any appreciable portion of its length. The compressive strength of cement is influenced by the same factors as the tensile strength.

The shearing strength of neat cement is about the same as the tensile strength, and it depends upon the same factors. Very little information regarding the shearing strength of neat cement is available.

38. Time of Set.—One of the most important properties of Portland cement is its property of setting and hardening, which is caused principally by the hydration of its three major constituents, tricalcium silicate, dicalcium silicate, and tricalcium aluminate. When water is added to Portland cement, these compounds first form amorphous and later both crystalline and amorphous hydrated materials. The tricalcium aluminate sets and hardens very quickly, and the "initial" set of the cement is undoubtedly due to the hydration of this compound. The early hardness and cohesive strength of the cement are due to the hydration of the tricalcium aluminate and the tricalcium silicate. The further increase in strength is due to the further hydration of these two compounds as well as that of the dicalcium silicate. The tricalcium silicate is the most important cementing compound of the three. This setting and hardening will progress under water as well as in air.

The actual time of set is of much importance in some work. It is not desirable to have the set occur before the concrete is placed, neither is it desirable to have too long a time elapse before the cement sets, especially if the cement is to be placed under water. In general, the higher the temperature the quicker the set takes place. An excess of water will lengthen the time required. Cement sets slower in damp weather than in dry. An addition of gypsum or plaster of Paris up to about 3 per cent retards the set, while a larger addition of plaster of paris will tend to give the cement a "flash" set. The seasoning of the cement often affects the time of set, sometimes increasing and sometimes decreasing the length of time required.

The specifications require that the cement shall not develop initial set in less than 45 minutes when the Vicat needle is used or 60 minutes when the Gilmore needle is used. Final set shall be attained within 10 hours.

39. Fineness.—It has been determined that the final particles of cement are the ones which give the cement its cementing values. Fineness of grinding increases the strength of cement mortars, but not that of neat cement pastes. Fine grinding also increases the sand carrying capacity of the cement, shortens the time of set, and is thought to make the cement more sound.

The specifications for fineness of cement require that 78 per cent or more of the cement shall pass a standard 200-mesh sieve.

40. Specific Gravity.—The specific gravity test of Portland cement is not of much importance and is not made unless it is specifically ordered. A low specific gravity may be caused by adulteration in large amounts, but small amounts may not have enough effect to lower the specific gravity below 3.10. There is practically no relation between the degree of burning and the specific gravity. Seasoning tends to lower the specific gravity, due to the absorption of carbon dioxide and moisture from the air.

The specifications require that the specific gravity of Portland cement shall not be less than 3.10 (3.07 for white Portland cement). Should the test of cement as received fall below this requirement, a second test shall be made upon an ignited sample.

41. Uses of Portland Cement.—At present, Portland cement is used very much in structural work and it is rapidly replacing lime, natural cement, and other kinds of cements in this field. As a part of mortar it is used for stone and brick masonry and for finishing coats, etc. As a part of monolithic concrete it is used for all kinds of heavy masonry work such as foundations, dams, piers, footings, abutments, retaining walls, pavements, sidewalks, etc. As a part of reinforced concrete it is used in walls, buildings, floors, roofs, piles, bridges, tunnels, subways, ships, conduits, pipes, culverts, etc. Portland cement ranks next to steel and timber as a structural material at the present time and it will probably outrank timber in the near future. At present, cement concrete is not so reliable a structural material as steel or timber, due to the fact that not so much is known about concrete and also that unskilled men are often employed for selecting the aggregate and mixing and laying of the concrete. There is no doubt but that the use of Portland cement in structural work will be much more extensive in the future than it is at the present time.

CHAPTER IV

PORTLAND CEMENT MORTARS

A. DEFINITIONS AND MATERIALS

42. Definitions.—A Portland cement mortar is a mixture of Portland cement, fine aggregate (sand or its equivalent), and water.

Fine aggregates are particles of gravel, crushed stone, sands, or other materials which will pass a ¼ in. sieve.

Silt is sometimes defined as particles between 0.005 mm and 0.05 mm. in diameter; clay as particles less than 0.005 mm. in diameter; and loam as a mixture of any of the fine materials with organic matter, either animal or vegetable.

43. The Cement and the Water.—The cement should be a Portland cement capable of passing the standard specifications. On the work, the cement should be stored in a weather-tight building which will protect it from dampness, and so piled as to permit of ready inspection and sampling. Whenever practicable, each shipment of cement should be sampled and tested before being used.

The water used for Portland cement mortar should be free from oils, acids, alkalis, and organic matter (either animal or vegetable). The water should not contain any chemical in solution that would be harmful to the mortar. The presence of oil is easily detected by its surface film. Organic matter (usually of vegetable origin) can sometimes be detected by observing floating particles, or by turbidity, though chemical tests are often required. Tests of water for acidity or alkalinity can be made by means of litmus paper. If there is any doubt as to the suitability of the water for use, its effect on soundness, set, and strength of the mortar should be determined by tests.

44. Sand in General.—In mortars and concretes it is just as important to have a good sand as it is to have a good cement. Due to the progress in the manufacture of Portland cement, the quality of most Portland cements is such that there are more failures due to the use of a poor sand than to the use of a poor cement.

The sand should be composed of a hard siliceous material free from loam, clay, sticks, animal or vegetable matter, friable materials, etc.; and the particles of sand should be small enough to pass through a quarter inch sieve. The best sand, as to size, is one which contains both coarse and fine grains in such proportions that the percentage of voids will be a minimum. A coarse grained sand is usually better than a fine grained one. While a sand should preferably consist of hard silica grains, other minerals may be present without causing any bad effects. However, sands containing mica, hornblende, feldspar, and carbonate of lime are not durable and should not be used. The physical condition of a sand is of more importance than the chemical composition. A friable sand is worthless. A small percentage of finely divided clay or loam is not usually injurious.

Sands may be washed to remove dirt and like materials, but care should be taken not to wash away too much of the finer particles of the sand.

45. Properties of Sand.—Other things being equal, the smaller the percentage of voids, the better the sand for use with Portland cement. The percentage of voids in dry sand ranges from about 25 to 45 per cent. The percentage of voids in a sand may be found by dropping a known volume of well-shaken dry sand into water and noting the volume displaced. The difference between the original volume of the sand and the volume of the water displaced gives the volume of voids. Or, as the specific gravity is nearly a constant (2.65) for all sands, the percentage of voids can be approximately determined from the weight per cubic foot. Another way is to pour water on the sand (contained in a water-tight vessel) until the surfaces of the sand and water coincide. The volume of water required is equal to the volume of voids in that amount of sand.

Well-shaken dry sand will weigh from 90 to 125 lb. per cubic foot, but if the sand is in a loose condition it may weigh as much as 20 per cent less.

Moist sand, that is not packed, weighs less than dry sand.

The percentage of absorption of sand rarely exceeds 3 per cent.

The specific gravity of sand is usually between 2.6 and 2.7 and the average value is about 2.65.

46. Sieve Analysis of Sand.—The sieve analysis of sand (or fine aggregate) is one of the best tests for determining the suitability of the sand for use in a Portland cement mortar. This analysis consists of sifting a sample of sand through several different sieves (five or more) and noting the amount passing each sieve. Sieve openings 1/10 in. or larger are usually in the form of circular holes, while woven brass wire cloth is used for the sieves with smaller openings. These woven-wire sieves are known by numbers corresponding to the number of openings per lineal inch. For analyzing sands the following sieves are desirable:

Sieve	Diameter of opening, inches	Sieve	Diameter of opening, inches	Sieve	Diameter of opening, inches
3/8 in.	0.375	No. 10	0.073	No. 40	0.015
1/4 in.	0.250	No. 15	0.046	No. 50	0.011
1/6 in.	0.167	No. 20	0.034	No. 80	0.007
1/10 in.	0.100	No. 30	0.022	No. 100	0.0055

The results of a sieve analysis may be shown graphically by plotting the sieve openings as abscissæ and the corresponding percentages passing each sieve as ordinates. This will give a curve from which the qualities of the sand may be estimated (see chapter on "Plain Concrete" for sample curves for sand).

The uniformity coefficient is the ratio of the diameter of the particles, represented by the point where the curve crosses the 60 per cent line, to the diameter of the particles where the curve crosses the 10 per cent line. A coarse sand has a uniformity coefficient of about 5.2 or more; a medium sand of about 4.2; and a fine sand of about 2.2. A sand that has a uniformity coefficient of about 4.5 is usually considered good for concrete work.

47. Standard Sand.—Standard sand is the sand recommended for use in cement testing by the American Society of Civil Engineers and other engineering societies. It is a natural bank sand obtained from Ottawa, Ill., U. S. A., and screened to proper size. Only the sand that passes a No. 20 sieve and which is held on a No. 30 sieve is used. The percentage of voids in this sand is about 37 per cent, and the weight per cubic foot is about 104 lb.

48. Substitutes for Sand.—Stone screenings are the fine materials (less than one-quarter of an inch in size) which have been screened out from crushed stone. When they are free from clay and dirt, they make a good substitute for sand. They are apt to be a little coarser than sand, but they have about the

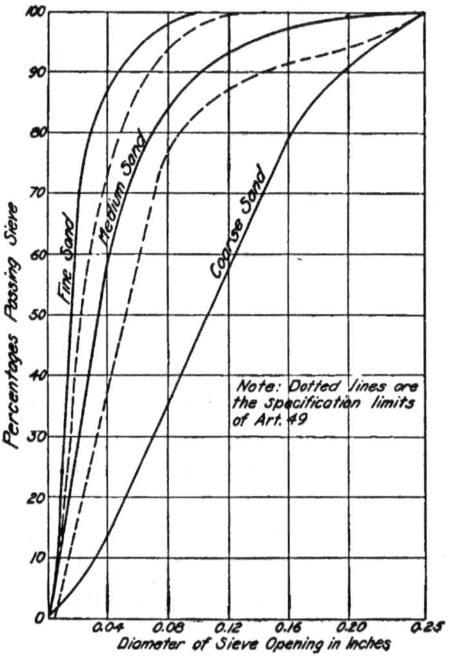

FIG. 16.—Mechanical analysis of sands.

same percentage of voids and weight per cubic foot. Screenings make a strong mortar, but the strength usually decreases more rapidly with a decrease in the amount of cement than in the case of a sand mortar.

Well selected and screened mine tailings often make as good a cement mortar as stone screenings. Granulated blast-furnace slag, small cinders, clay, loam, etc. have been used as substitutes for sand in a Portland cement mortar, but they do not make so good a mortar as ordinary sand.

49. Specifications for Fine Aggregate.—These specifications are practically the same as those adopted by the New York Public Service Commission. Fine aggregates for use in a Portland cement mortar or concrete should conform to the following requirements.

SIEVE ANALYSIS OR MECHANICAL GRADING

Sieve	Diameter of opening, inches	Per cent passing sieve limits
¼ in.	0.250	100
$^3/_{16}$ in.	0.187	Between 93 and 100
No. 6	0.138	Between 90 and 100
No. 10	0.073	Between 75 and 93
No. 15	0.047	Between 48 and 80
No. 30	0.022	Between 20 and 50
No. 50	0.011	Between 2 and 30
No. 100	0.0055	Between 0 and 7

Curves for the extreme conditions may be plotted upon cross-section paper together with the curve for the sand tested. If the sand is good, its curve will lie between the two extreme curves. In general, a sand for use in a mortar may be a little coarser than a sand for use in a concrete.

Silt.—Not over 7 per cent of the dry weight of the sample should pass a No. 100 sieve when screened dry.

Strength.—Both tensile and compressive strengths of **a 1:3** mortar (proportioned by weight) shall be equal to or more than the strengths required for a 1:3 standard Ottawa sand mortar in the standard and tentative (proposed) specifications of the American Society for Testing Materials.

Organic Matter.—The loss on ignition shall not exceed $^1/_{10}$ of 1 per cent of the total dry weight.

B. PROPORTIONING AND MIXING MORTAR

50. Proportioning the Mortar.—The proportioning of cement and sand for a mortar is usually done by one of the three following methods: (1) by weight; (2) by volumes of packed cement and loose sand; and (3) by volumes of loose cement and loose sand.

The best way of proportioning the materials is by weight, and this method is usually followed in the laboratories, though rarely in practical work. The presence of moisture in the sand may affect the proportioning to some extent if the amount of moisture is not approximately determined and allowed for. Sand rarely contains more than 5 per cent of moisture by weight; hence, the error due to moisture would usually be less than 5 per cent if no correction were made.

Proportioning by packed cement and loose sand is probably the second best method. This method is used to some extent on practical work. Usually a sack of cement is considered to be 1 cu. ft. in volume and only the sand is measured. In measuring the sand, it is important to secure the same degree of looseness each time, otherwise the proportions may be changed. The difference in volume between loose and compact material may be as much as 20 per cent in some cases.

Proportioning by loose cement and loose sand measured by volume is the least reliable of the three methods due to the inability or neglect of the average workman to secure the same degree of compactness at all times. The common method is to dump the cement and sand loosely into measuring boxes and then empty the boxes on the mixing platform. Often the measuring is done by pails or wheelbarrows and the proportioning is very inaccurately done.

The proportions of mortar for masonry work are usually a 1:2 or a 1:3 mix, 1 part of cement to 2 or 3 parts of sand. Sometimes as rich a mix as a 1:1 is required for finishing or other work while as lean a mix as a 1:5 may be used in some cases.

51. Mixing the Mortar.—The mixing should be done either by hand or by machine. In either method it is better first to mix the cement and sand dry and in the proper proportions, and then add the water and mix again. The batches should not be too large.

Machine mixing is faster than hand mixing and the quality is more uniform. The cement and sand are first placed in the machine and mixed for a minute or so. Then the water is added and the batch mixed for a few minutes more. A well-handled mixer will turn out a batch of mortar every 5 minutes.

For hand mixing, suitable water-tight platforms must be provided to prevent the loss of cement. The sand is first spread out in a layer on the platform and the cement is then placed in a thin layer on top of the sand. The cement and sand are then mixed dry until they are of a uniform color. Then the water is added and the batch is thoroughly mixed again. Shovels and hoes are convenient tools to use in the mixing. Thoroughness of mixing is of the most importance.

The mortar should be used before the initial set has taken place. Cement mortar that has reached initial set should not be used. Retempering (remixing) of cement mortar should not be allowed after the initial set is reached.

C. PROPERTIES OF PORTLAND CEMENT MORTARS

52. Strength of Portland Cement Mortars in General.—The strength of Portland cement mortar depends upon (1) the proportion of cement used; (2) the size and grading of the sand; (3) the amount of water used; and (4) the degree of compactness of the mortar. That is, the strength of the mortar depends upon (*a*) the amount of cement per unit volume, and (*b*) the density of the mortar.

In order to secure the best results, it is necessary to make tests upon different mixes of cement, sand, and water. Uniform conditions of testing must be carefully observed in order to secure reliable results. The strength of Portland cement mortar is affected by the temperature of the air and water, the thoroughness of gaging, and the conditions of testing. It is necessary to standardize the methods of testing before trying to determine the influence of the constituents of the cement, sand, and water used in the mortar. This testing should preferably be done by experienced operators in a well-equipped laboratory. The personal equations of different operators will have some effect on the results and care should be taken to minimize this effect as much as possible (see any book on the testing of Portland cement and cement mortars for the standard methods).

53. Effect of Density and Size of Sand on the Strength.—Density of the mortar may be defined as the ratio of the actual solid material (absolute volume of the cement and sand) to the total volume of the hardened mortar. The density may be determined by carefully weighing the materials used and assuming a value of 3.1 for the specific gravity of the cement and 2.65 for the specific gravity of the sand. If the actual values for the specific gravities have been obtained, these values should be used.

In general, the size and grading of the sand that will give the densest mortar will also give the strongest mortar. This requires that the percentage of voids shall be small and that the sand shall have a sufficiency of coarse grains. A low percentage of voids depends upon the grading of the sand and not on the actual size of the grains. With the same percentage of voids, a coarse sand will make a stronger mortar than a fine sand. If a sand is not suitable for use, it may be made suitable by mixing another sand or part of another sand with it so as to give a low percentage of voids. Sometimes a sand may be screened in two or three different sizes and these sizes remixed in different proportions so as to reduce the amount of voids.

Feret made a study of the effect of the size of sand grains on the strength of Portland cement mortar and his results showed that (1) the densest mortar was generally the strongest; (2) the proportion of fine sand should be small; and (3) if the sand is uniform in size, a coarse sand is better than a medium sand and a medium sand is better than a fine sand.

54. Effect of the Amount of Mixing Water on the Strength.—An increase in the amount of water used (above the proper amount needed) in the mixing of the Portland cement mortar will (1) increase the time required for setting; (2) decrease the strength of neat cement mortar, having a greater effect on short time than on long time tests; (3) decrease the strength of the mortar on short time tests (say under 6 months), but will have less effect on the results of long time tests; (4) increase the amount of laitance on the surface; (5) increase the difficulty of bonding the new mortar to the old; and (6) tend to cause a segregation of the materials (sand and cement).

A decrease in the amount of water (below the proper amount required) used in mixing the mortar will (1) tend to hasten the set; (2) increase the voids; (3) decrease the strength, except that a slight decrease in the amount of water used may increase the strength, especially on short time tests, provided that the mortar is well compacted; and (4) make the mortar less water-tight.

55. Effect of Various Conditions on the Properties of Mortars. Mortars made and used in dry weather should have their exposed surfaces kept moist for several days so that the water will not be evaporated from these surfaces before the mortar has hardened. Portland cement mortar will harden a little more rapidly in dry weather.

Mortars made and used in wet weather require a little more time to set and attain their strength.

Hot weather (high temperatures) decreases the time required for set and increases the rate of gain in strength.

Low temperatures increase the length of time required for the setting and hardening of the cement mortar and decrease the strength on short time tests. At a temperature of 40 degrees Fahrenheit, the strength is only about two-thirds of that at 70 degrees Fahrenheit when the mortar is 2 months old. Cement requires about four times as long to set at a temperature of 32 degrees Fahrenheit as it does at a temperature of 65 degrees Fahrenheit.

Freezing of Portland cement mortars retards their rate of hardening and their rate of increase in strength. Exposed surfaces, that are frozen before the final set occurs, often scale off. It is not good practice to use Portland cement mortar in freezing weather unless special precautions are taken to keep it from freezing.

Regaging or remixing a Portland cement mortar after setting has begun is generally not permitted. Experiments on the effect of regaging mortars gave various results, depending upon different mortars and the length of time elapsed between the mixings. Regaging of some mortars within a few hours after the initial set had taken place caused no bad results whatever, but in most cases such regaging seemed to cause a decrease in the ability to harden as well as a decrease in the strength. It is thought that the effect of regaging a Portland cement mortar within 2 hours after mixing is not very injurious.

56. Effect of Various Elements **on the Properties** of **Mortars.** A small percentage (2 or 3 per cent) of mica added to a 1:3 Portland cement mortar may cause a 20 per cent loss of strength due to an increase in voids and the inability of the cement to stick to the smooth surface of the mica.

Dirt has an injurious effect on the strength of the mortar, especially if it contains any organic matter.

As small as 1/10 of 1 per cent of organic matter may be injurious.

A small percentage of any friable material is injurious.

Clay usually decreases the strength of the mortar, but in some cases a small amount of finely divided clay (say from 5 to 10 per cent), which has been thoroughly mixed with the sand, appears to have a good effect. A rich mortar is generally injured by the addition of clay while a lean mortar may be improved, especially if the mortar contains a large percentage of voids.

Good finely divided loam has about the same effect as clay.

Lime has about the same effect as clay but is not thought to be so injurious. Small percentages of lime often improve a lean mortar but may injure a rich mortar. A small addition of lime paste makes a cement mortar much easier to work with in laying brick or stone masonry.

Salt, when added to the mixing water, lowers the freezing point of the mortar and, up to about 10 per cent, appears to have but little effect on the strength. For temperatures below 32 degrees Fahrenheit, the amount of salt required to lower the freezing temperature 1 degree Fahrenheit is about 1 per cent of the weight of the mixing water.

57. Tensile Strength.—The statements made in preceding paragraphs on the strength of neat Portland cement apply equally well to the strength of a Portland cement mortar. In determining the qualities of a cement that is to be used, the tensile strength of the mortar is more valuable than the tensile strength of the neat cement. Under normal conditions the strength of Portland cement mortar increases very rapidly during the first few days. The rate of gain of strength gradually decreases. At the age of 7 days, the strength is about one-half or two-thirds of the maximum, which is reached at an age of about 3 months. The specifications (minimum requirements) for a 1:3 standard sand mortar are as follows:

One Part of Portland Cement to 3 Parts of Standard Ottawa Sand

Age and Storage	Tensile Strength, Pounds Per Square Inch
1 day in moist air, and 6 days in water........	200
1 day in moist air, and 27 days in water........	300

The proportions are by weight. The temperature of the materials during the mixing, storing, and testing should be as near 70 degrees Fahrenheit as practicable.

A good sand and cement should give results much higher than the above minimum requirements for a standard sand mortar.

Many mortars show a slight retrogression in strength after 5 or 6 months, but this retrogression is usually not permanent and it does not appear at all in the compression test results.

58. Compressive Strength.—Testing a Portland cement mortar in compression is the best way of judging of the suitability of the cement and sand for construction purposes. In general, a mortar that is strong in tension is also strong in compression, but the ratio of the strengths is not a constant quantity. The compressive strength of a good mortar increases steadily with age and shows no retrogression.

The modulus of elasticity in compression is a variable quantity because the stress strain curve is not a straight line. At about one-fourth of the ultimate strength, the modulus of elasticity in compression is approximately 4,000,000 lb. per square inch for neat cement and about 3,000,000 lb. per square inch for a good 1:3 mortar.

At present there are no standard specifications for the compressive strength of Portland cement mortar in America, but the following minimum requirements have been proposed and adopted as tentative specifications:

One Part of Portland Cement to 3 Parts of Standard Ottawa Sand

Age and Storage	Compressive Strength, Pounds Per Square Inch
1 day in moist air, and 6 days in water	1,200
1 day in moist air, and 27 days in water	2,000

The specimens are cylinders 2 in. in diameter and 4 in. high. Each value should be the average of not less than three specimens, and the average at 28 days must be higher than that at 7 days.

A good mortar should give results that are much higher than the above minimum requirements for a standard sand mortar.

59. Transverse Strength.—The transverse strength of a Portland cement mortar, as calculated from the formula $S = Mv/I$, is approximately two times the tensile strength. The cross-bending strength is proportional to the tensile strength, and it depends upon the same factors.

60. Adhesive Strength.—The adhesive strength of neat Portland cement and Portland cement mortars, at the age of 6 months, with a few different materials, is shown by the following table:

Adhesive Strength in Pounds per Square Inch

Mixture	Iron Rods	Sawn Limestone	Brick
Neat	315	270	50
1:1	290	220	40
1:2	265	170	30
1:3	110	75	15

61. Shearing Strength.—The shearing strength is of importance, as concretes and mortars are often subjected to shearing stresses in practical work. Shearing tests are rarely ever made on account of difficulties of obtaining a true shearing stress. The shearing strength depends upon the same factors as the tensile and compressive strengths. The shearing strength is usually proportional to the compressive strength. The fineness of grinding of the cement and the qualities of the sand are the most important factors.

The following table gives the results of some tests upon the shearing strength of neat Portland cement and Portland cement mortars made by Bauschinger in 1879. The specimens were about 2½ by 5 in. in cross-section and were stored in water. Each result is an average of nine tests. It is to be noted that the cement used did not pass the tension test requirements of the American specifications; the neat strength being 224 lb. per square inch for the 7-day and 294 lb. per square inch for the 28-day tests, while the 1:3 mortar results were 95 lb. per square inch for the 7-day and 169 lb. per square inch for the 28-day tests.

Shearing Strength in Pounds per Square Inch

Mix	Age 7 Days	Age 28 Days	Age 2 Years
Neat	271	346	415
1:3	116	188	375
1:5	77	131	364

62. Miscellaneous Properties.—*Abrasive* resistance of Portland cement mortars depends not only upon the cement but also upon the hardness of the sand grains.

Expansion and Contraction.—Cement mortar, when hardening, in air, will contract slightly, and when hardening in water it will keep a nearly constant volume or expand a little. The richer the mortar, the greater the effects of expansion and contraction.

Permeability is the measure of the rate of flow of water through a mortar of a given thickness and under a given pressure. An impermeable mortar is a water-tight one. Permeability decreases rapidly for all mixtures with an increase in the age of the specimens tested; it decreases considerably with a continuation of flow; and it increases with an increase of pressure, leanness of mix, dryness of mixture, and with increased coarseness of the sand used. An addition of a small amount of finely divided clay or loam tends to decrease the permeability. (See the articles on "Impervious Concrete" in the chapter on "Plain Concrete" for methods and materials for decreasing the permeability of plain concrete. These articles apply to a Portland cement mortar as well as to plain concrete.)

Absorption of water by a Portland cement mortar depends upon the same conditions (but not to so large an extent) as the permeability does. In general, the absorption decreases slightly with age; increases with an increased leanness of mixture; and the dry mixtures are slightly more absorptive than the wet ones.

Voids in a Portland cement mortar depend upon the grading of the materials and the consistency of the mix. A well-graded aggregate with a small percentage of voids will usually give a mortar with a small percentage of voids. If more or less water is used than is necessary to form a proper consistency, the voids in the mortar will be increased. A dry mix usually has more voids than a corresponding wet mix. The voids in a mortar usually vary between 15 and 30 per cent.

Weight.—A good Portland cement mortar of a 1:3 mix will weigh about 140 lb. per cubic foot; a 1:1 mix about 145 lb.; and a 1:4 mix about 138 lb. per cubic foot. The weight per cubic foot varies directly with the density of the mortar and the specific gravities of the cement and fine aggregate used.

CHAPTER V

PLAIN CONCRETE

A. DEFINITIONS AND MATERIALS

63. Definitions.—Concrete is an artificial stone made by mixing cement, water, and an aggregate consisting of large and small particles, such as broken stone or gravel and sand or screenings.

Aggregates are those inert materials which, when bound together by cement, form a concrete.

Fine aggregate is usually defined as the material that will pass a ¼-in. sieve, while coarse aggregate is the material which is held on a ¼-in. sieve.

64. Cement, Water, and Fine Aggregate.—*Cement.*—The cement used should preferably be a Portland cement that will pass the standard specifications of the American Society for Testing Materials (or equivalent specifications) when subjected to the standard tests recommended by the American Society of Civil Engineers.

After delivery at the work, the cement should be carefully stored in weatherproof buildings having tight floors above the ground level in order to protect the cement from the weather and to allow of ample time for inspection and testing. If kept dry, the cement will not be injured by a long storage and it may be improved, due to the seasoning. Before being used, each shipment of cement should be carefully inspected, sampled, and tested by a competent person. In sampling, one sample should be taken from about every tenth barrel and care should be taken to secure a fair sample. The amount of cement required for the standard tests is about 10 lb.

Water.—The water used in making concrete should be clean and free from any impurities which would be injurious to the concrete. See the discussion regarding a suitable water for Portland cement mortars in the preceding chapter (Chap. IV, Art. 43).

Fine Aggregate.—The fine aggregate used in making concrete should be a good sand, or its equivalent, and should possess those requisites that are given and discussed in the preceding chapter (Chap. IV) on "Portland Cement Mortars." In general a sand for use in a concrete should possess more fine particles than a sand for use in a Portland cement mortar.

The fine aggregate should be stored in bins or piles convenient to the work and, if necessary, be screened to remove large particles and be washed to remove dirt and silt. In washing, care should be taken not to wash out too much of the finer material.

65. Coarse Aggregate in General.—The coarse aggregate used for concrete usually consists of crushed stone, gravel, cinders, slag, broken brick, etc. Any stone is suitable for concrete work that is durable and strong enough so that the strength of the concrete will not be limited by the strength of the stone. Strength, density, hardness, toughness, durability, and cleanliness are desirable properties in a coarse aggregate. As the physical character of a rock depends upon its mineral constituents and structure, only those rocks which have durable mineral constituents and a dense, strong structure should be used for concrete. Rocks which are structurally weak or which contain weak mineral constituents should not be used. Granites, traps, and limestones are often employed for concrete work, while sandstones are rarely suitable for this work. Soft, flat, or elongated particles do not make a satisfactory material for use in concrete. Clean screened gravel is a good substitute for broken stone, but it often contains some particles of a soft friable nature that will reduce the strength of the concrete. Cinders, and sometimes slag, may be used for a coarse aggregate for a concrete subjected to very low stresses or which may be used as a fireproofing material or where light weight is desired. Broken brick should not be used in concrete work of any importance or where strength is required.

After the stone is quarried, it may be broken by laborers with stone hammers or it may be crushed in stone crushers. Jaw crushers are usually used in small or portable plants and gyratory crushers in large stationary plants Screening of the crushed stone or gravel is often necessary to remove the dust and other fine material that will pass a ¼-in. sieve. Sometimes it is necessary to wash the gravel to remove the dirt, loam, clay, or organic matter adhering to it.

Coarse aggregate may be stored in bins or piled in the open without any special protection from the weather. Care should be taken to keep the coarse aggregate clean and prevent dirt, clay, loam, organic matter, etc. from being mixed with it.

66. Size of Coarse Aggregate.—The maximum size of crushed stone for concrete work varies according to the use to which the concrete is to be put. When crushed stone is used for massive walls, the maximum size may be 2½ or 3 in.; 2 in. for abutments; 1½ in. for arch rings; 1 in. for copings, bridge seats, and thin walls; and 1 in. or ¾ in. for reinforced concrete work. Flat, irregular, or rough stones are not so desirable as are the more rounded ones.

A crushed stone or gravel that is nearly all of one size is not so good as an aggregate that is made up of uniformly graded particles because an aggregate all of one size usually has a larger percentage of voids. It is often desirable to screen the aggregate into two or more sizes and remix these sizes in different proportions in order to secure the proper grading.

A mechanical (sieve) analysis is of value in studying the grading of a coarse aggregate that is to be used in concrete work. The sieves used are preferably ones of 2½-, 2-, 1¾-, 1½-, 1¼-, 1-, ¾-, ½-, ⅜-, and ¼-in. mesh. Usually all of these sieves are not needed for any one test, but just a sufficient number should be used to give the desired information. The results of a sieve analysis may be plotted on cross-section paper and a curve drawn through the points (see article on "Proportioning by Mechanical Analysis" for further discussion).

67. Voids, Weight per Cubic Foot, and Specific Gravity of Coarse Aggregates.—The voids in a coarse aggregate may be found by pouring a known volume of the aggregate into a known volume of water and noting the displacement. In measuring the volume of the aggregate, care must be taken to secure a uniform degree of compactness. The volume of the aggregate minus this displaced volume equals the volume of the voids.

Another way of determining the voids in a coarse aggregate is to pour water on a known volume of the aggregate, contained in a water-tight vessel, until the surfaces of the aggregate and the water coincide. The volume of the water added equals the volume of the voids.

The percentage of voids varies from 30 to 55 per cent for common crushed stone and gravel, depending to some extent on the shape, grading, and degree of compactness.

The weight per cubic foot for coarse aggregate (crushed stone and gravel) usually varies from about 75 to 120 lb. Crushed stone is often sold by the cubic yard but it is frequently measured by weight, 2,500 lb. being considered equal to 1 cu. yd.

The specific gravity of stone and gravel varies somewhat. Approximate values are as follows: trap 2.8 to 3.0; granite 2.65 to 2.75; limestone 2.6 to 2.7; sandstone 2.3 to 2.6; and ordinary sand and gravel 2.6 to 2.7.

The following table shows the relation between voids, weight per cubic foot, and specific gravity:

VOIDS AND WEIGHT OF BROKEN STONE AND GRAVEL

Percentage of voids	Weight in pounds per cubic foot				
	Specific gravity 2.6	Specific gravity 2.7	Specific gravity 2.8	Specific gravity 2.9	Specific gravity 3.0
35	106	110	114	118	122
40	97	101	105	109	112
45	89	93	96	100	103
50	81	84	87	91	94
55	73	76	79	82	84

68. Specifications for Coarse Aggregate.—The following specifications are those of the American Railway Engineering and Maintenance of Way Association:

Stone shall be round, hard, and durable, and shall be crushed to sizes not exceeding 2 in. in any direction. For reinforced concrete, sizes usually are not to exceed ¾ in. in any one direction, but the size may be varied to suit the character of the reinforcing materials.

Gravel shall be composed of clean pebbles of hard and durable stone of sizes not exceeding 2 in. in diameter, and shall be free from clay and other impurities except sand. When the gravel contains sand in any considerable quantity, the amount of sand per unit of volume of the gravel shall be determined accurately, to admit of the proper proportion of sand being maintained in the concrete mixture.

The following specifications for coarse aggregate for concrete are practically the same as those adopted by the New York Public Service Commission.

Cleanliness.—All broken stone aggregate must be so free from dust that the limit of fineness (5 per cent) shall not be exceeded (fine material being the material passing the ¼-in. sieve). All gravel must be thoroughly washed, preferably at the plant or pit where it is secured.

Mechanical Grading.—A sieve analysis shall be made of the coarse aggregate and, if the aggregate is suitable for use, the results should be within the limits given in the following table. If so desired, curves for the extreme conditions or limits for the kind of sieves used may be plotted on cross-section paper together with the curve for the coarse aggregate tested. If the coarse aggregate is good for use, its curve will lie between the two extreme or limiting curves.

MECHANICAL GRADING

SIZE OF OPENING, INCHES	SQUARE-HOLED SIEVES, LIMITS, PERCENTAGE PASSING	ROUND-HOLED SIEVES, LIMITS, PERCENTAGE PASSING
2	100	100
1½	Between 95 and 100	Between 75 and 95
1¼	Between 65 and 92	Between 50 and 85
1	Between 40 and 80	Between 35 and 70
¾	Between 25 and 60	Between 20 and 50
½	Between 10 and 40	Between 7 and 35
¼	Between 0 and 5	Between 0 and 5

B. PROPORTIONING OF CONCRETE

69. General Theory.—The theory of proportioning is that the fine and the coarse materials should be so proportioned that the concrete will have the greatest density. This means that the voids in the concrete should be a minimum. This is accomplished when there is just enough cement to fill the voids in and completely coat all of the particles of the sand, and just enough mortar to fill all the voids in and completely coat all of the particles of the coarse aggregate. There must be no excess of water, otherwise water voids will be formed.

Proportioning by weight will secure more uniform mixtures than proportioning by volume because the errors due to the measurement of the materials in a loose or compact form are eliminated. These errors may be as large as 20 per cent. In practical work the proportioning is nearly always done by volume because this method is more convenient.

70. Proportioning by Standard Proportions.—Proportioning in practical work is commonly done by "rule of thumb," using certain standard proportions. The materials are measured by volume, the unit of measurement being 1 cu. ft. usually. The following are some of the standard mixes:

1:1:2	A very rich mixture used only where great strength and water tightness are required.
1:1½:3	A rich mixture not quite so strong as the first, but used for the same purposes.
1:2:4	A good mixture used very often in reinforced concrete work and for foundations subjected to vibrations.
1:2½:5	A medium mixture used for floors, retaining walls, abutments, etc.
1:3:6	A lean mixture used for massive concrete structures under steady loads of not great intensity.
1:4:8	A very lean mixture used only for massive concrete work which is not very important.

71. Proportioning with Reference to Coarse Aggregate.—The theory of this method is that just enough mortar should be used to fill the voids in the coarse aggregate. In practice, more mortar is required, because of the separation of the coarse aggregate by the mortar and excess water and the consequent increase in the voids. About 10 per cent more mortar is required on an average. If care is taken to secure a properly graded coarse aggregate, the voids will be less, and less mortar will be required to produce a concrete of the required strength and imperviousness. This means a saving of cement and sand. In general, ordinary proportioning by voids is no better than arbitrary proportioning, because of the behavior of the different materials when mixed together to form a concrete.

72. Proportioning with Reference to Mixed Aggregate.—The theory of this method is to grade both the coarse and the fine aggregates together so as to reduce the percentage of voids in the mixture to a minimum. Then the amount of cement required will depend upon the strength and imperviousness desired. The amount of cement necessary to fill completely the voids of the mixture may be estimated by making a void test on a well shaken mixture of the aggregates. In practical work it has been found that slightly more cement is required to make a concrete of maximum density because of the slight increase in voids formed when the cement and water are added. This method is no better than the preceding one.

73. Proportioning by Maximum Density Tests.—Different mixtures of fine and coarse aggregates may be mixed with the required amounts of cement and water and the resulting concrete tested to determine which proportions of fine and coarse aggregates are the best.

The procedure of the test is roughly as follows: A trial mix of fine and coarse aggregates is prepared, the proper amounts of cement and water are added, and the whole thoroughly mixed. The resulting concrete is placed in a water-tight metal cylinder and tamped. Then the volume of the concrete is measured. Other trial mixes of the fine and the coarse aggregates are prepared and the volumes of the concretes formed are carefully measured. The same amounts of cement and water should be used each time and the total weight of all materials and water should be the same for all of the tests.

The mixture which gives the least volume is the best mixture and will make the strongest, densest, and most impervious concrete.

Care should be taken not to use too much water when making the tests as the excess water will increase the voids in the concrete and thus destroy the accuracy of the tests.

The test described above is usually called a "yield" test on concrete.

74. Proportioning by Mechanical Analysis.—This is a good and accurate method of properly proportioning the concrete materials. A sieve analysis is first made of each of the aggregates and the results plotted on cross-section paper, using the percentages passing a given sieve as ordinates and the corresponding sizes of sieve openings as abscissæ. A curve is drawn for each aggregate. It is not necessary to draw a curve for the cement as it completely passes practically all of the sieves.

By using the curves, the materials can be so proportioned (by cut and try methods) that they will give a mechanical analysis curve that agrees very closely with the ideal curve, or curve of maximum density. This ideal curve consists of a portion of an elliptic curve and a straight line. The straight line is drawn from the intersection of the maximum size of the coarse aggregate and the 100 per cent lines tangent to an elliptical curve. The ordinate of this point of tangency is equal to 33 per cent, and the abscissa is equal to $\frac{1}{10}$ of the maximum size of the coarse aggregate. The elliptical curve is drawn from this point of tangency to the origin. Aggregates of apparently unsuitable grading may be studied in this way and the proper proportions determined. Sometimes it is found necessary to screen a coarse aggregate into two or more sizes and then to combine these sizes in different proportions in order to obtain a dense mixture.

75. Example of Proportioning by Mechanical Analysis.—Suppose that for a 1:9 concrete, it is desired to find the proper

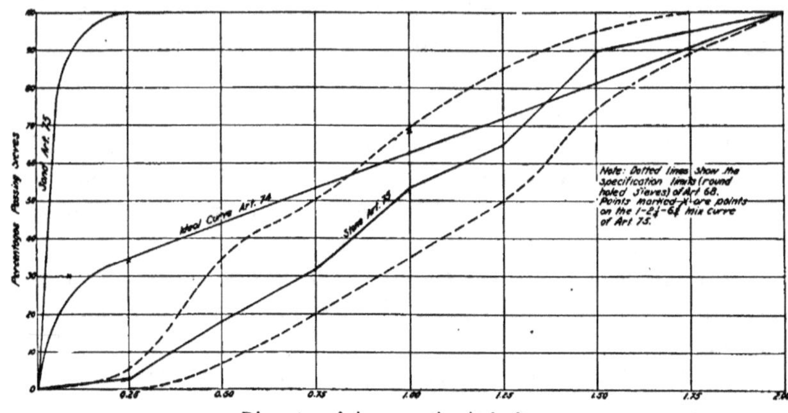

Fig. 17.—Mechanical analyses curves, etc.

proportions of sand and stone whose sieve analysis gave the following results:

Sand		Stone	
Sieve	Per cent passing	Sieve	Per cent passing
¼ in.	100	2 in.	100
No. 6	95	1½ in.	90
No. 10	86	1¼ in.	65
No. 15	66	1 in.	54
No. 30	34	¾ in.	32
No. 50	17	½ in.	18
No. 100	5	¼ in.	3

The sieve-analysis curves for the aggregates and the ideal curve as directed in the preceding article should be plotted. A tabulation similar to the following should then be made:

Concrete 1:9 Mix Per Cent Passing Sieves

Material	No. 30	No. 15	No. 10	No. 6	¼ in.	½ in.	¾ in.	1 in.	1¼ in.	1½ in.	2 in.
Cement..	10	10	10	10	10	10	10	10	10	10	10
Sand.....	..	..	19½	..	22½	22½	22½	22½	22½	22½	22½
Stone...	..	..	½	..	3	....	....	36½	....	....	67½
Total	..	..	30	..	35½	....	....	69	....	....	100

Because of lack of uniformity in the grading of the materials and possible errors in the sieve analysis, computations of the percentages to the nearest ½ per cent will be of more than sufficient accuracy for the problem.

The proportion of cement in a 1:9 mix is $\frac{1}{10}$ or 10 per cent of the whole, and as all of the cement will pass all of the sieves in the tabulation 10 per cent may be written for all values for the cement.

To obtain the proportions for the first trial curve, the percentage where the ideal curve crosses the ¼-in. sieve opening line should be noted. (This is about 35 per cent in this case.) Then the sum of the cement, sand, and stone passing this sieve should equal about 35 per cent. As the cement is 10 per cent (see tabulation), 25 per cent is left for the sand and stone. As all of the sand passes the ¼-in. sieve and only 3 per cent of the stone, 22.5 per cent may be taken for the sand. Then the proportions for the first trial curve will be 1 part (10 per cent) cement, 2¼ parts (22.5 per cent) sand, and 6¾ parts (67.5 per cent) stone.

Now, as all of the sand passes the ¼-in. and larger sieves, 22.5 per cent may be written in the tabulation for the values of the sand passing these sieves.

The amount of stone passing the ¼-in. sieve (expressed as a percentage of the dry materials) is determined by taking 67.5 per cent of 3 per cent (see sieve analysis), which gives about 2 per cent.

The total amount of dry materials passing the ¼-in. sieve is 34.5 per cent.

Suppose that the total percentage passing the 1-in. sieve is next determined. There will be 10 per cent cement, 22.5 per cent sand (see tabulation), and 67.5 per cent of 54 per cent (see sieve analysis) or 36.5 per cent stone, giving a total of 69 per cent.

Suppose that the total percentage passing the No. 10 sieve is determined next. There will be 10 per cent cement (see tabulation), 22.5 per cent of 86 per cent (see sieve analysis) or 19.5 per cent sand, and 67.5 per cent of about 1 per cent or approximately 0.5 per cent stone, giving a total of 30 per cent. (For the smaller sieves, no harm will be done if the percentage of stone passing is neglected as the values will be less than 0.5 per cent in this problem.)

In the same way, the total percentages passing all of the other sieves should be determined and recorded in the tabulation.

Then all of these total percentages should be plotted and a smooth curve drawn through the points.

If the trial curve agrees very closely with the ideal curve, the proportions chosen are correct.

If the trial curve is mostly above the ideal curve, the proportion of sand should be decreased and the proportion of stone increased and a new trial curve computed and plotted.

If the trial curve is mostly below the ideal curve, the proportion of sand should be increased and the proportion of stone decreased and a new trial curve computed and plotted.

The process should be continued until a satisfactory curve is obtained or the materials are found to be unsuitable as they are. An experienced operator rarely has to make more than three trial curves.

For a good working concrete, the total percentages passing the smaller sieves should not fall below the ideal curve, as it is better to have a slight excess of fine material than to have too little of it. In regard to the total percentages passing the larger sieves, it is immaterial whether the trial curve is a little above or a little below the ideal curve.

A complete solution of the above problem should be made so that the method may be thoroughly understood.

76. Proportioning of Concrete Mixes by Abrams' Method.[1]—This method is probably the most scientific method of concrete proportioning yet proposed as it is based on fairly definite relations between the sieve analysis of the aggregate, the consistency of the mix, and the strength of the concrete. It is frequently called the "fineness modulus method."

The fundamental principle of this method is that, other things remaining the same, the quantity of mixing water used determines the strength of the concrete as long as the mix is plastic. Professor Abrams found that the compressive strength of a concrete made with a certain aggregate depended on the ratio of the volume of water to the volume of cement in the mix. He also found that there was a close relation between the size and grading of the aggregate, as measured by its fineness modulus, and the

[1] NOTE.—Practically all of this article is taken, with Professor Abrams' permission, from his bulletin on the "Design of Concrete Mixtures" published by the Structural Materials Research Laboratory, Lewis Institute, Chicago.

amount of water required to produce mixes of uniform plasticity or consistency.

The fineness modulus is a term used to denote the ***effective grading*** of an aggregate. It is computed from the sieve analysis of the aggregate. For making the sieve analysis the following

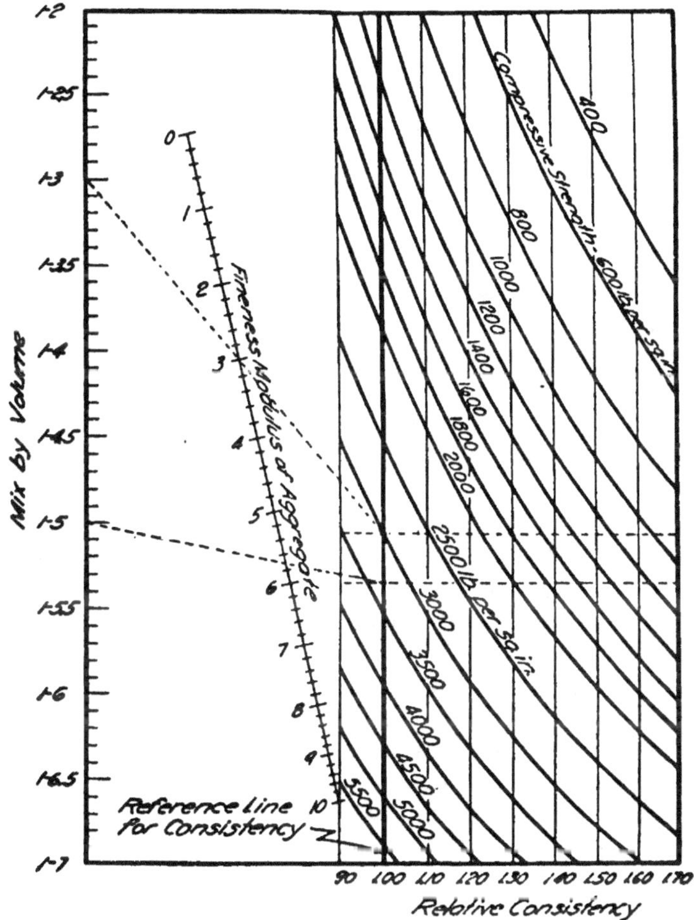

FIG. 18.—Abrams' chart for the design of concrete mixers.

Tyler standard sieves are used: Nos. 100, 48, 28, 14, 8, 4; and ⅜ in., ¾ in., and 1½ in. Each sieve has a clear opening just double that of the preceding one. The percentages of the sample *coarser* than (held on) the sieves are tabulated. The ***fineness modulus*** of the aggregate is the sum of the percentages divided by 100. The coarser the aggregate, the higher the fineness modulus.

Results of tests showed that mixtures of given fine and coarse aggregates having the same fineness modulus and the same amounts of cement and water produced concretes of equal strength (provided that the concrete was plastic and the aggregates were not too coarse for the amount of cement used) and of equal consistency. Tests also showed that the compressive strength of concrete mixes increased with the fineness modulus up to a certain value, after which the strength decreased. This point of maximum strength occurred at higher values of the fineness modulus for rich mixes than for lean mixes.

The following procedure of proportioning by fineness modulus is practically the same as that proposed by Professor Abrams:

1. Make sieve analysis of the fine and the coarse aggregates, using Tyler standard sieves of the following sizes: Nos. 100, 48, 28, 14, 8, 4; and ⅜ in., ¾ in., and 1½ in. Express the sieve analysis in terms of percentages by weight of material *coarser* than each of the standard sieves.

2. Compute fineness modulus of each aggregate by adding the percentages obtained in the sieve analysis and dividing by 100.

3. Determine the maximum size of the aggregate by applying the following rules: If more than 20 per cent of the aggregate is coarser than any sieve, the maximum size shall be taken as the next larger sieve in the standard set; if between 11 per cent and 20 per cent is coarser than any sieve, the maximum size shall be the next larger "half sieve;" if less than 10 per cent is coarser than certain sieves, the smallest of these sieve sizes shall be considered the maximum size.

4. Assume a mix and from Professor Abrams' table determine the maximum size of fineness modulus which may be used for the mix, kind and size of aggregate, and the work under construction.

5. Compute the percentages of fine and coarse aggregates required to produce the fineness modulus desired for the final aggregate by applying the formula: $P = \frac{(A - B)}{(A - C)} 100$

when P = percentage of fine aggregate in total mixture.
A = fineness modulus of coarse aggregate.
B = fineness modulus of final aggregate mixture.
C = fineness modulus of fine aggregate.

6. Using Professor Abrams' chart, draw a straight line through the mix and fineness modulus of aggregate used. Note where this line intersects the reference line of consistency (relative consistency of 1.00). Through this point of intersection draw a horizontal line. Read the strength at the point where this horizontal line crosses the vertical line for the relative consistency considered in design (see following paragraphs in regard to consistency).

7. If the mix chosen does not give the desired strength, select another mix and try again until a suitable mix is found.

8. Whenever time permits, make test cylinders of the mixes selected and test them at age of 28 days for a check on the design of mix.

The use of Professor Abrams' chart is a great help in approximately determining the mix required for a certain compressive strength when the relative consistency and fineness modulus are known.

The relative consistency of 1.00 is about as low as can ever be used, and this consistency requires tamping. For ordinary plain concrete work, a relative consistency of about 1.10 is as low as should be used for designing. For ordinary reinforced concrete work, a relative consistency of about 1.20 is advised for use in designing.

In making concrete the least amount of mixing water that will produce a mix of workable consistency should always be used.

The compressive strength values given in Professor Abrams' chart were determined from compression tests on 6 by 12-in. cylinders stored in a damp place and tested at an age of 28 days. The values obtained in practical work will probably be lower than these because of: less care used in mixing, handling, and placing of concrete; variations in quality of cement; variations in consistency; different storage or curing conditions; and variations in age.

Consequently, whenever practicable, test cylinders should be made and stored under conditions similar to that prevailing on the work and then tested at an age of 28 days as a check on the design of the mix.

Maximum Practical Values of Fineness Modulus (Abrams)

Size mix of aggregate	Properties by volume				Aggregate: Cement			
	1	2	3	4	5	6	7	9
Mortars								
0–14	3.00	2.70	2.50	2.30	2.15	2.05	1.95	1.85
0–. 8	3.80	3.40	3.10	2.90	2.75	2.65	2.55	2.45
0– 4	4.75	4.20	3.90	3.60	3.45	3.30	3.20	3.05
Concretes								
0–⅜	5.60	5.05	4.70	4.40	4.20	4.05	3.95	3.85
0–½*	6.05	5.45	5.10	4.80	4.60	4.45	4.35	4.25
0–¾	6.50	5.90	5.50	5.20	5.00	4.85	4.75	4.65
0–1*	6.90	6.30	5.90	5.60	5.40	5.25	5.15	5.00
0–1½	7.35	6.70	6.30	6.00	5.80	5.65	5.55	5.40
0–2*	7.75	7.10	6.70	6.40	6.20	6.05	5.95	5.80
0–3	8.20	7.55	7.15	6.85	6.60	6.50	6.40	6.25

* Half sieves not used in computing fineness modulus.

For *mixes* other than those given in the table, use the values for the next leaner mix.

For *maximum sizes* of aggregate other than those given in the table, use the values for the next smaller size.

This table is based on the requirements for *sand-and-pebble* or *gravel* aggregate, composed of approximately spherical particles, in ordinary uses of concrete in reinforced concrete structures. For other materials and in other classes of work, the maximum permissible values of fineness modulus for an aggregate of a given size is subject to the following corrections:

1. If *crushed stone* or *slag* is used as coarse aggregate, reduce values in table by 0.25. For crushed material consisting of unusually flat or elongated particles, *reduce* values by 0.40.
2. For *pebbles* consisting of *flat particles*, *reduce* values by 0.25.
3. If *stone screenings* are used as fine aggregate, *reduce* values by 0.25.
4. For the top course in *concrete roads*, or other work requiring a smooth finish, *reduce* the values by 0.25. If finishing is done by *mechanical means*, this reduction need not be made.
5. In work of *massive proportions*, such that the smallest dimension is larger than 10 times the maximum size of the coarse aggregate, additions may be made to the values in the table as follows: for ¾-in. aggregate, 0.10; for 1½-in., 0.20; for 3-in., 0.30; for 6-in., 0.40.

Sands with fineness modulus lower than 1.50 are undesirable as fine aggregate in ordinary concrete mixes. Natural sands of such fineness are seldom found.

Sand or screenings used for fine aggregate in concrete must not have a higher fineness modulus than that permitted for mortars of the same mix. Mortar mixes are covered by the table and by (3) above.

Crushed stone mixed with both finer sand and coarser pebbles requires no reduction in fineness modulus, provided the quantity of crushed stone is less than 30 per cent of the total volume of the aggregate.

77. Proportioning Concrete by Edwards' Surface Area Method. This method is one proposed by Mr. L. N. Edwards. From the results of tests Mr. Edwards found that, with aggregates of uniform quality, the strength of the mortar or concrete depended on (1) the amount of cement used in relation to surface area of the aggregate, and (2) the consistency of the mix. He also found that, other things being equal, the fine aggregate with a small total surface area gave a mortar of greater strength than a like aggregate having a greater surface area. Also that the amount of water required to produce a mortar of normal consistency depends on the amount of cement used and the total surface area of the fine aggregate wetted.

The general method of procedure for proportioning concrete by this method is as follows:

1. Make a sieve analysis of the aggregate.
2. Find average number of particles per unit weight of the aggregate passing one sieve and held on another.
3. From the results of (2) and the specific gravity of the particles, compute the average volume of each size of particle.
4. Compute the surface areas from the average volumes of the various sizes and shapes of the particles. (Grains of sand and gravel were assumed as spherical, while particles of broken stone were assumed to be one-third cubes and two-thirds parallelopipeds.)
5. Determine the total surface area of the aggregate.
6. Base the quantity of cement on the total surface area.
7. Base the quantity of water on the quantity of cement and the total surface area of the aggregate.
8. Make strength tests on the mortar or concrete as determined in (7).
9. Increase or decrease the cement and water content of the mix until a mix is found that gives the strength required.

The proper water-cement ratio must always be maintained or else the strength results will not be satisfactory.

The work required for this method of proportioning can be simplified in the laboratory by the use of curves and tables showing relations between surface areas and unit weights of particles of various sizes and specific gravities, water-cement ratios, relations between strength and cement content and surface areas, etc.

78. Formula for Estimating Quantities of Materials Required for Plain Concrete.—The following formula is of use in determining the quantities of cement, fine and coarse aggregates required for a certain volume of concrete. The units may be any convenient units of volume, provided that, in any one problem, the units of the concrete, cement, fine and coarse aggregates are the same. Customary units are cubic feet or cubic yards (and often barrels for the cement).

The number of parts of cement required are $\frac{1.55 \times C}{C + S + R}$.

The number of parts of fine aggregate (sand) required are $\frac{1.55 \times S}{C + S + R}$.

The number of parts of coarse aggregate (stone) required are $\frac{1.55 \times R}{C + S + R}$.

C, S, and R are the proportions of cement, sand, and stone, respectively, in the mixture. The proportions are by volume.

The number of parts of a material obtained from the formula must be multiplied by the volume of the concrete to obtain the volume of the material required. If the volume of the concrete is in cubic feet, the volumes of the cement, sand, and stone will be in cubic feet also.

To reduce cubic feet of cement to barrels, divide by 4 (assuming 4 cu. ft. per barrel). Some authorities assume 3.8 cu. ft. per barrel and use 3.8 as a divisor instead of 4.

To reduce cubic feet of sand or stone to cubic yards, divide by 27, there being 27 cu. ft. in 1 cu. yd.

When the volume of concrete is small enough to be made in one batch, it is customary to increase slightly the quantities of materials required to allow for waste due to the concrete sticking to the mixing boards, machines, barrows, tools, etc. This increase in the quantities is usually about 5 or 10 per cent.

C. MIXING OF CONCRETE

79. Hand Mixing.—As the strength of concrete depends to a large extent upon the thoroughness of mixing, great care should be taken in mixing. The mixing should be done on watertight platforms of sufficient size to accommodate the men and materials for the progressive and rapid mixing of at least two batches of concrete. These batches should be small, not exceeding 1 cu. yd.

The sand should be evenly spread upon the platform and the cement on top of the sand. The cement and sand should then be mixed dry until the mixture is of a uniform color. Enough water should be added to make a thin mortar, and the materials

FIG. 19.—Concrete mixer. (*Koehring Machine Co.*)

FIG. 19a.—Concrete mixer.

mixed again. Then the coarse aggregate, which has been thoroughly wetted, should be placed on top of the mortar and the whole batch thoroughly mixed until it is of a uniform consistency. More water may be added, if necessary, so that the batch will be of the desired consistency. From three to five turns are required at each stage of the mixing.

Some engineers prefer to mix the cement, sand, and stone dry and then add the water, and thoroughly mix again.

80. Machine Mixing.—Machine mixing is usually much better and quicker than hand mixing and should generally be required when the amount of work is enough to make machine mixing economical. In machine mixing all of the materials (including the water) are usually introduced at once without any intermediate mixing. However, some authorities think that better results will be obtained by first mixing the dry materials and then adding the water and mixing again. The time required for mixing depends upon the type and speed of the mixer, and varies for different machines. For the best results, the time of mixing should rarely be less than 1 minute after the water is added. Machine mixed concrete is often stronger than hand mixed concrete.

Machine mixers are of two kinds, the batch and the continuous type. In a batch mixer, the proper amounts of materials for one batch are added and mixed and discharged from the mixer, and then the operation is repeated again. The mixing is done by moving paddles or blades or by the rotation of the receptacle itself. The batch mixer usually consists of a fixed or revolving drum with movable or fixed paddles or blades inside. In a continuous mixer the operation of mixing is practically continuous, care being taken to maintain the proper proportions of the materials. These mixers usually consist of a trough containing some form of screws or paddle wheels to assist in the mixing. The batch mixer gives better results because it is easier to supervise the operations and also to secure the proper proportions of the materials in the concrete.

81. Consistency of Concrete.—At the present time most engineers favor a consistency where just enough water has been used so that the concrete will just flow. This consistency gives a concrete that can be deposited by means of spouts and pipes and which requires but little puddling or tamping in the forms to make it homogeneous and to secure smooth surfaces next to the forms.

A consistency that requires tamping to make the concrete quake is usually better, stronger, and more impervious than the above consistency, but it is not quite so economical in placing and tamping.

A very wet concrete contains more voids and is weaker than the above concretes and, if it is not quickly placed, there is a tendency for the materials to segregate.

A dry concrete requires much tamping and very careful inspection in order to secure good work.

D. DEPOSITION OF CONCRETE

82. Forms for Concrete.—In general, the forms used for concrete should be durable and rigid and well braced to prevent any bulging or twisting, as well as strong and tight enough to prevent leakage of the concrete.

The forms for concrete are usually constructed out of cheap rough lumber, such as rough pine or spruce, using the better grades of lumber only when a very smooth concrete surface is desired. Green lumber is better than dry because it will not be affected by the water in the concrete to so large an extent. The forms should be constructed so that, after the concrete has hardened, they can be readily removed without much damage to the lumber. Lumber for ordinary forms can be used from three to five times on an average before it is damaged enough to be thrown away.

Sometimes metal forms are used. There are a number of types of these forms on the market. A good metal form can be used very many times before it becomes worn out. Consequently, metal forms are economical when the amount of work will admit of their repeated use, even though their first cost is greater than that of wooden forms.

83. Transporting, Placing, and Tamping Concrete.—In general, the transportation system must be such that the concrete will be carried from the mixer to the forms before it attains initial set; that no part of the concrete will be lost in transporting; that no segregation of materials will take place; that the delivery of the concrete will be continuous and uninterrupted; and that the transporting will be done efficiently, rapidly, and economically. Some of the methods of transportation are shovels, wheelbarrows, carts, large buckets, cableways, pipes, spouts, spouting plants (including hoists, dump buckets, concrete bins, pipes, and spouts), etc.

In placing concrete care should be taken to see that the concrete shall be continuously and evenly placed; segregation avoided; laitance and stoppage planes prevented; voids reduced; and lateral flow prevented. Also that concrete, that has been remixed after initial set has taken place, should not be placed in the forms. Extra care is required in depositing concrete in cold or very hot weather. Usually, the concrete is placed in the forms in layers about 6 or 8 in. deep and a new layer added before the other has set.

All concrete should be puddled or tamped when it is placed in the forms to eliminate voids, bring any free water to the surface, secure a close filling of the forms and contact with the reinforcement, make the concrete more homogeneous, and make a dense mortar coat and smooth finish at the exterior surfaces.

After placing the concrete, care should be taken to prevent a too rapid drying. It should be kept moist for at least 2 weeks after the removal of the forms.

84. Bonding New Concrete to Old Work.—In joining fresh concrete to hardened concrete or to old masonry, the surface of contact should be thoroughly cleaned of all loose material, dirt, and laitance before depositing the fresh concrete. If a strong bond is desired, the surface should be washed with a dilute acid solution and water and then plastered with a coat of rich cement mortar or grout. If necessary, the contact surface should be roughened with tools, as a good bond can be more easily secured on a rough surface.

A cement grout is a very thin and, usually, rich mortar.

Laitance is a whitish, chalky substance washed out of the cement in the concrete. This substance appears on the surface of practically all concretes, has little or no strength and hardening properties, weakens the bond between old and new concrete, and spoils the appearance of the structure. It may be removed by scrubbing.

85. Surface Finish of Concrete.—As the surface of concrete often shows the marks of the forms after their removal, it is sometimes necessary to finish the surface to improve its appearance.

One method is to finish the surface by tooling. The use of different tools (chisels, hammers, etc.) will give a variety of finished surfaces. Changing the angle at which the chisel is held will also change the appearance of the surface.

Another method is to give the surface a rub finish by rubbing it with carborundum stone, emery, concrete, or a soft natural stone. This type of finish is not expensive and gives a good surface.

Sometimes the forms covering the exposed surfaces are removed and the concrete surface brushed while the concrete is still green. A wire brush is usually used.

Sandblasting is sometimes used to finish a concrete surface that is thoroughly hardened. Care should be taken not to make depressions in the surface or to round off the edges too much.

A very pleasing finish is secured by the use of colored aggregates in the concrete and then properly finishing the surfaces.

White cement, pigments, and stains, etc. are often used to secure desired surface finishes.

Plaster coatings should rarely ever be applied to concrete surfaces as they are generally not durable.

86. Placing Concrete Under Water.—Concrete may be placed under water if there is little or no current flowing, as a current of water tends to wash the cement out of the concrete before it can harden.

Probably the best way of placing concrete under water is by using a metal tube (called a "tremie") about 1 ft. in diameter, slightly flaring at the bottom, and long enough to reach from the bottom to above the surface of the water. The tube should be kept full of concrete all the time, and the bottom of the tube should be moved about slowly so as to allow of the gradual discharge of the concrete.

Another method is to deposit the concrete in large quantities in fairly tight molds and then not disturb it before it attains its final set.

Bottom dumping buckets fitted with top covers have been used more or less successfully in some places.

Sometimes the concrete is placed in bags of loosely woven cloth, jute, or burlap and then deposited under the water.

87. Placing Concrete in Freezing Weather.—If it can be avoided, concrete should not be placed in freezing weather. However, fairly good work may be done if proper precautions are taken to keep the concrete from freezing before it sets. This may be accomplished in various ways such as heating the forms by steam pipes, by adding salt to the mixing water, by heating the materials before mixing and placing them, or by some combination of these methods. If salt is added to the mixing water, it requires about 1 per cent of salt by weight to reduce the freezing temperature of the concrete 1 degree Fahrenheit. More than 8 or 10 per cent of salt should never be used (except in extreme cases) as it may seriously lower the strength of the concrete. Calcium chloride has less effect on the strength of the concrete than the ordinary sodium chloride has. Also, calcium chloride accelerates, while sodium chloride retards, the setting of the concrete.

Alternate freezing and thawing of concrete have a bad effect on the strength and should be very carefully guarded against. The freezing of concrete after the final set has taken place does but very little damage to it.

E. IMPERVIOUS CONCRETE

88. Impervious Concrete in General.—While it is impossible to make concrete actually waterproof, it may be made practically impervious by several different methods or combinations of these methods. Though it is practically impossible to keep out all of the water, yet the water may be prevented from passing through the concrete in such amounts as to cause inconvenience and damage. In general, the flow of water through the concrete varies directly with the amount of voids in the concrete (the voids may be large, due to imperfect grading of the aggregates, excess of mixing water, improper mixing, improper placing, segregation of materials, etc.), the water pressure or head on the concrete, the amount of laitance, and the number of shrinkage or temperature cracks. The flow of water through the concrete varies inversely with the age, the density, and the amount of cement.

89. Effect of Increasing the Density and Amount of Cement.—Concrete may be made more water-tight by making it more dense, but this requires great care in the proportioning, mixing, and placing. The more dense the concrete, the more impervious it is. Greater density may be secured by a proper grading of the materials so as to secure a minimum percentage of voids. An excess of mixing water must be avoided, as this excess will form water voids. Also, too little water must not be used as a dry concrete is not so impervious as a slightly wet mix. A concrete made by using a slightly wet consistency with well-graded aggregates and which is thoroughly puddled in the molds will give a fairly water-tight structure. Care must be taken to prevent cracks.

Increasing the amount of cement used tends to make the concrete more water-tight. The richer the mix, the better it is. A concrete leaner than a 1:6 should not be used.

Increasing the density and the amount of cement are the best methods in use at present for increasing the water-tightness of concrete.

90. Using **Waterproofing Materials.**—Concrete may be made practically water-tight by placing layers of a waterproofing material between layers of the concrete, but great care must be taken in order to secure a good bond between these layers. Sometimes a layer of asphalt is placed between two layers of concrete. Often a tar paper or a roofing felt is placed between concrete layers and bonded to them by a coating of hot asphalt or tar. Frequently several layers are used. Continuity of the paper or felt is important

91. Using Foreign **Matter in the Concrete.**—An addition of from 8 to 15 per cent of hydrated lime (based on the weight of the cement) to the concrete aids in reducing the porosity and thus making the concrete more water-tight. Other materials such as fireclay, feldspar, ground sand, etc. have been used for the same purpose with varying success.

Sometimes small percentages (2 or 3 per cent) of an alum soap solution (1 part alum to 2.2 parts soap), chloride of lime, oil emulsions, and similar compounds have been used to make the concrete more impervious by acting as a void filler and a water repellant. As most of these compounds tend to weaken the concrete large percentages should not be used.

92. Use of **Surface Treatments.**—In order to make the concrete more impervious, the surface may be given one or more coats of oil paint, varnish, bitumen (asphalt, petroleum, coal tar, etc.), a paraffin solution in benzine or benzol, soap, soap and alum, cement grout, or cement mixed with a waterproofing material. A plastering with a rich cement mortar is effective, provided a good bond can be secured and cracks can be prevented from forming. Surface treatments aid in making existing structures more water-tight.

F. PROPERTIES OF CONCRETE

93. Effect of Various Impurities Mixed with the Concrete.—*Mica.*—A very small amount of mica in concrete will cause an appreciable loss of strength.

Dirt.—An appreciable amount of dirt is apt to have an injurious effect on the strength of concrete.

Organic Matter.—All organic matter should be carefully excluded from concrete. As small an amount as $\frac{1}{10}$ of 1 per cent may be injurious.

Clay.—A small amount of clay is sometimes beneficial, especially in lean mixes, but a large amount (over 10 per cent) will weaken the concrete. A concrete that has a large percentage of voids may be improved by the addition of a small amount of finely divided clay.

Loam.—Loam has about the same effect as clay.

Lime.—Unhydrated (quicklime) lime should never be placed in concrete as the expansion during the hydration will probably cause the disintegration of the concrete. Hydrated lime has about the same effect as clay. Small percentages of hydrated lime often improve porous or lean concretes but are usually injurious to rich or dense concretes. In general, hydrated lime is preferable to clay.

Sugar.—A very small percentage of sugar has a bad effect on the strength and soundness of concrete.

Grease and Oil.—These materials have an injurious effect on the concrete when mixed with the concrete materials in the making of the concrete.

Regaging.—Regaging, retempering, or remixing a portland cement concrete is generally not permitted as it is thought to be very injurious to the concrete. Concrete that has passed the stage of initial set should never be placed in the forms.

94. Effect of Various Elements on Hardened Concrete.—*Fire.* Concrete has better fire-resisting qualities than ordinary brick, stone, tile, or terra cotta. Concrete may be heated to 1,200 degrees Fahrenheit (this is as hot as an ordinary fire) for 3 or 4 hours and then suddenly cooled with a stream from a fire hose without showing more than a slight surface disintegration. A thickness of about 2 in. of concrete over steel is enough to keep the steel from warping, bending, twisting, or being otherwise damaged in an ordinary building fire. Of course, aggregates which will ignite or disintegrate at comparatively low temperatures (less than 1,700 degrees Fahrenheit) should not be permitted in a concrete that may be subjected to fire.

Acids.—A thoroughly hardened concrete of first-class quality is affected only by strong acids such as would seriously injure other materials.

Grease and Oil.—A concrete, properly made and hardened, and with a carefully finished surface, will resist the action of ordinary engine oils and petroleum.

Sea Water.—Sea water has practically no effect on good concrete made with properly manufactured cement and good aggregates and containing a very small percentage of voids. Some concrete structures exposed to sea water have been seriously disintegrated. This action usually occurs at the water line and may be shown by a swelling and cracking of the concrete, a softening and crumbling of the mortar, or a formation of a crust which later cracks off. The disintegration of reinforced concrete is often due to the sea water penetrating the concrete and causing the steel to rust. In most cases the disintegration is due to poor cement, poor aggregates, or poor proportioning, mixing, or placing of the concrete. A rich mortar coat, a few inches thick, applied to the concrete at the water line tends to protect the concrete from the action of the sea water. It is thought that the active element in the sea water is sulphate of magnesia.

Alkali.—It has been frequently noted that concrete, which has been partly submerged in alkali water, often disintegrates near the water line. The disintegration is like that of concrete exposed to sea water. Just what chemical action takes place is not known, but the active element is thought to be sulphate of magnesia. Aggregates which contain alkali should not be used in concrete work.

95. Effect of Varying the Amount of Mixing Water.—Water has four functions in concrete; *i.e.*, (1) to react with the cement and form a binding material; (2) to aid in spreading the cement over the surfaces of the aggregates; (3) to act as a lubricant between the particles of the fine and the coarse aggregates and thus aid in the placing of the mixture in the molds; and (4) to occupy space in the concrete.

There should be enough water present to react with all of the cement. If too little water is used, the reaction will not be complete; while if too much water is used, the mixture will be too dilute to develop the best strength. These effects are shown by strength tests on concretes of dry, normal, and wet consistencies. While a dry mix may give greater strength in short time tests, the normal mix is nearly always the stronger after a few months have elapsed.

Enough water should be present to carry the particles of cement and spread them over the surfaces of the particles of the aggregates. Too little water will not allow of the proper spreading of the cement particles, while too much water will tend to keep the cement from sticking to the aggregates.

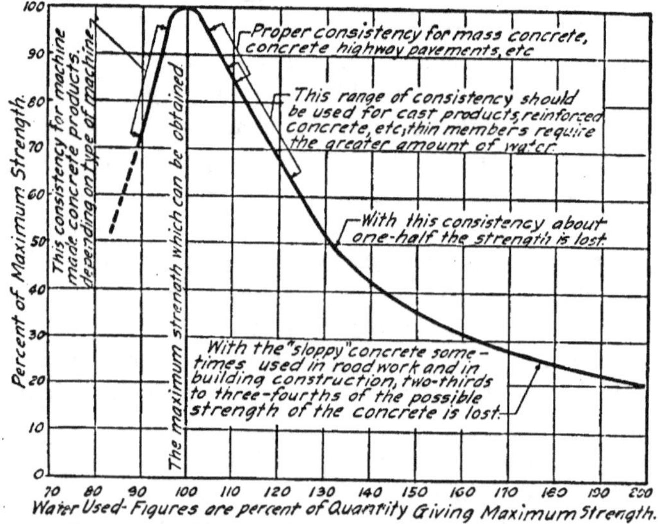

FIG. 20.—Effect of quantity of mixing water on strength of concrete. (*Abrams.*)

Too little water retards the flowing of the concrete and makes it difficult to place the concrete in the molds and compact it properly, while too much water tends to cause segregation of the materials. Enough water should be present for the proper lubrication of the materials and no more.

Water occupies space in the concrete and, if too much is used, it tends to push the solid particles farther apart and make the mixture less dense. Further, this excess of water may escape, after the concrete has set, and leave air voids.

Excess of water also has the following bad effects on concrete: (1) it tends to cause day-work planes; (2) it tends to cause large deposits of laitance; (3) it makes the concrete less impervious; (4) it increases the difficulty of bonding new to old concrete;

(5) it tends to make dusty concrete floor surfaces; and (6) it increases the difficulties of concreting in freezing weather.

It is very important that just the proper amount of water be used in concrete work and that the engineer in charge of the work regulate the water at all times so as to secure the proper consistency. Different cements and different aggregates (and sometimes different batches of the same aggregates) often require slightly varying amounts of water for the normal consistency.

96. Strength of Concrete in General.—In general, the strength of a Portland cement concrete depends upon: (1) the amount of cement per unit volume; (2) the density of the concrete; and (3) in some cases upon the strength of the aggregates. Of course, the strength of concrete increases with its age, but the rate of increase decreases with the age.

Any factors which influence any of the above conditions will also affect the strength of the concrete. Some of these factors are: the consistency; the conditions of mixing, placing, and storing or aging; the qualities of the cement, water, fine and coarse aggregates; presence of impurities; etc.

Results of tests indicate that concretes stored or aged in damp or moist air are stronger than those aged in water or dry air. Also, that concretes exposed to the weather (sun, wind, and rain) are usually stronger than concretes cured indoors in a comparatively dry room.

A concrete of a slightly dry consistency, well mixed and thoroughly tamped, is generally stronger than a concrete of slightly wet consistency, but the wetter consistency gives better results in practical work and is necessary in reinforced concrete work. Also, a concrete of a slightly wet consistency becomes about as strong as the dry mix at the age of 6 months. A very wet mix or a very dry mix never becomes so strong as a normal mix and should not be used if it can be avoided.

With good grading, the actual size of the stone has but little effect on the strength of the concrete. Usually, a small size of stone is less well graded and gives less density when mixed with the sand. For plain concrete, the maximum size of the stone should rarely be less than 1 in. The maximum size in reinforced concrete work depends upon the molds and the spacing of the reinforcement.

Tests show that broken stone generally makes a stronger

concrete than gravel, though this difference is not very great (about 10 per cent).

The strength of the coarse aggregate may have an effect on the strength of the concrete if enough cement is used so that the failure takes place in the aggregate. Ordinary stone and gravel have enough strength for most kinds of concrete. Soft, friable stones, such as some of the sandstones, will give a weaker concrete. Cinders, brick, old concrete, etc. should be carefully investigated as to their strength before being used in concrete.

The presence of such materials as will reduce the strength of neat cement and cement mortar will also tend to reduce the strength of the concrete.

97. Compressive Strength of Concrete.—The compressive strength of concrete depends primarily upon the amount of cement per unit volume and also upon other conditions such as were discussed in the preceding article.

For a 1:2:4 mix of concrete, made under reasonably good conditions as to the character of the materials and workmanship, an average strength of 2,000 lb. per square inch may be expected at the age of 1 or 2 months. Under similar conditions, a 1:3:6 mix should average about 1,600 lb. per square inch. Poorer or better results may be obtained, depending upon the quality of the materials and the workmanship.

An average of 25 cylinders, each 10 in. in diameter and 24 in. long, of a 1:2:4 mix of machine made concrete tested at the University of Wisconsin gave an average strength of 1,940 lb. per square inch at an age of 30 days. An average of 44 cylinders of the same kind gave a strength of 2,150 lb. per square inch at an age of 60 days. A fairly fine sand was used. The consistency of the concrete was soft, and the specimens were stored in air and kept moist by sprinkling.

The following table gives results of tests made on 12-in. cubes at the Watertown Arsenal. Standard Portland cement, a clean coarse sharp sand, and crushed stone (having a maximum size of 2½ in. and 49.5 per cent of voids) were used. The cubes were buried in wet ground after their removal from the molds.

The compressive strength of concrete increases with age, reaching about 80 or 90 per cent of its ultimate at the age of 2 months.

The compressive strength of a good cinder concrete is about one-third of the strength of a corresponding mix of a good stone concrete.

Compressive Strength of Twelve Inch Cubes of Concrete

Mix	Strength in pounds per square inch			
	Age 7 days	Age 1 month	Age 3 months	Age 6 months
1:2:4	1,565	2,399	2,896	3,826
1:3:6	1,311	2,164	2,522	3,088

From the results of tests it has been observed that the strength of short concrete columns (as long as 10 or 15 diameters) is from 10 to 20 per cent less than that of short concrete prisms.

98. Tensile Strength of Concrete.—Satisfactory tensile tests of concrete are very difficult to make. The tensile strength varies from about 1/10 to 1/12 of the compressive strength. The quality of the materials and the workmanship both have a very great effect on the tensile strength. The same factors that affect the compressive strength also affect the tensile strength. The tensile strength of a well-made concrete at an age of 60 days is about as follows:

1:2:4 mix of concrete........ 175 to 275 lb. per square inch
1:3:6 mix of concrete........ 125 to 200 lb. per square inch

99. Transverse Strength of Concrete.—The transverse strength of concrete depends upon the tensile strength. The computed modulus of rupture is about twice the tensile strength and from 1/5 to 1/6 of the compressive strength. The following table gives an idea of the cross-bending strength of good concrete of various mixes at an age of one month:

Transverse Strength of Concrete (on Tension Side) 1 Month Old

Mix of concrete	Modulus of rupture, pounds per square inch	Mix of concrete	Modulus of rupture, pounds per square inch
1:1½:3	475	1:3:5	275
1:2:4	425	1:3:6	225
1:2:5	350	1:4:8	125

100. Shearing Strength of Concrete.—The shearing strength of concrete is of importance especially in short concrete columns and reinforced beams. Satisfactory shearing tests on concretes are hard to make, due to the difficulty of securing apparatus that will give a pure shearing stress. The shearing strength of concrete usually varies from ½ to ⅔ of the compressive strength. Tests on concrete at the University of Illinois gave the following results. The shear specimens were restrained beams and the compression specimens were cubes. The specimens were stored in damp sand.

Mixture	Shear, pounds per square inch	Compression, pounds per square inch	Ratio $\frac{\text{Comp.}}{\text{Shear}}$	Ratio $\frac{\text{Shear}}{\text{Comp.}}$
1:2:4	1,418	3,210	2.26	0.44
1:3:6	1,313	2,428	1.85	0.54
1:3:6	1,020	1,721	1.69	0.59

101. Adhesive Strength of Concrete to Steel.—The adhesive strength (or bond) of concrete to steel is of great importance in reinforced concrete work. This strength depends upon the richness of the mix and on the character of the surface of the steel. Corrugated rods usually give greater bond stresses than plain rods. Bond tests have been made in two different ways. One way was to measure the force required to pull a rod out of a block of concrete (pull out test), and the other method was to determine the force required to make a rod slip in a beam. The tests showed that the bond between the concrete and steel was divided into two parts; the adhesion between the concrete and steel and the sliding resistance. The adhesive strength may be said to have been reached when the first end slip of the rod was observed. This stress is about ½ or ⅔ the maximum stress attained. The beam tests are thought to have given more reliable values than the pull out tests. Results of beam tests gave maximum bond stresses varying from 160 to 375 lb. per square inch for round rods while square and flat rods were not quite so strong. Corrugated bars gave higher results. The concrete was a 1:2:4 mix. Pull out tests on specimens of the same mix usually gave higher results. There does not seem to be any relation between the size of rod and the unit bond stress. The bond strength of a 1:3:6 concrete is about 20 or 30 per cent less than that of a 1:2:4 mix.

102. Elastic Limit and Modulus of Elasticity of Concrete.—As the stress strain curve for concrete is not a straight line throughout any part of its length and as the concrete is subject to a permanent deformation even for a small load, concrete may be said to have no true elastic limit. There appears to be a limit, however, to the stress that can be repeated indefinitely without continuing to add appreciably to the deformation. This limit may be taken as the elastic limit, or yield point, for all practical purposes. From the results of tests it appears that this limit is usually somewhere between 40 and 60 per cent of the ultimate.

As the stress strain curve for concrete is a curved line, the modulus of elasticity is not a constant through any appreciable range of stress. One way to determine the modulus of elasticity is to take the slope of the curve at the origin. Another, and perhaps a better, way is to compute the secant modulus for a load of 300 or 500 lb. per square inch, or for a load equal to about ⅓ of the ultimate. The second way usually gives a value considerably less than that obtained by the first.

For a concrete one month old and for a stress of 500 lb. per square inch, the secant modulus of elasticity for a 1:2:4 mix will generally be between 2,000,000 and 2,500,000 lb. per square inch, and between 1,500,000 and 2,000,000 lb. per square inch for a 1:3:6 mix. If the modulus of elasticity is computed from the slope of the curve at the origin, the value obtained will probably be from 20 to 50 per cent higher than the secant modulus at 500 lb. per square inch.

In general, the modulus of elasticity increases with the richness of mix and the age, but varies greatly with different aggregates.

103. Yield of Concrete.—Yield may be defined as the volume of concrete that may be obtained from given quantities of cement, fine and coarse aggregates. Other things being equal, those concrete materials should be used which will give the greatest yield of concrete. This means that less quantities of the materials will be required for a given volume of concrete and, consequently, the cost of the concrete will be less, assuming that the materials are purchased by volume and that the prices are the same for all varieties of the same kind of materials.

6

104. Expansion and Contraction of Concrete.—Experiments have shown that concrete will shrink a little when hardening in air, and that when it is hardening under water it will keep about the same volume or perhaps swell a trifle. The coefficient of expansion for concrete is about 0.000006 per degree Fahrenheit. The coefficient of expansion increases but very little with an increase in the richness of the mix. The fact that an average crushed stone concrete has a coefficient of expansion practically equal to that of steel is of importance in reinforced concrete work.

105. Miscellaneous Properties of Concrete.—*Weight per Cubic Foot.*—The weight per cubic foot of concrete may vary considerably, due to the kind of materials used for aggregates. The weight also varies directly with the richness of mix and the density. A concrete made from sand and crushed stone usually weighs from 135 to 160 lb. per cubic foot. For practical purposes, the weight of concrete may be assumed to be 145 or 150 lb. per cubic foot.

Absorption.—The absorption of water by concrete may be quite small or very large, depending upon the richness and density of mix, kind of materials used for aggregates, thoroughness of mixing, care in placing, etc. In general, the same factors that tend to make concrete impervious will also tend to make it non-absorptive.

Abrasion.—The abrasive resistance of a concrete depends primarily upon the abrasive resistance of the mortar. Of course, if the surface of the concrete is worn away so that the coarse aggregate is exposed, the abrasive resistance of the coarse aggregate will have some influence on the abrasive resistance of the concrete. The abrasive resistance of the mortar depends upon the ability of the cement to hold the sand grains together and also upon the abrasive resistance of the sand grains themselves.

106. Working Stresses and Factor of Safety for Concrete.—The following working stresses are recommended by the Committee on Concrete and Reinforced Concrete of the American Society of Civil Engineers.

The allowable compressive stress on a short plain concrete column or pier (whose length does not exceed 12 diameters) is 22.5 per cent of the strength at 28 days, or 450 lb. per square inch for 2,000 lb. concrete. The factor of safety is 4.5.

The extreme fiber stress in compression in a reinforced concrete beam, calculated on the assumption of a constant modulus of elasticity for concrete under working stresses, may be allowed to reach 32.5 per cent of the compressive strength at 28 days, or 650 lb. per square inch for 2,000 lb. concrete. The apparent factor of safety is 3.1 while the actual factor is larger.

Where pure shearing stress occurs, uncombined with compression normal to the shearing surface and with all tension normal to the shearing plane provided for by reinforcement, a shearing stress of 6 per cent of the compressive strength at 28 days, or 120 lb. per square inch for 2,000 lb. concrete, may be allowed. The factor of safety in this case is between 6 and 7.

When the shear is combined with an equal compression, as on a section of a column at 45 degrees with the axis, the stress may equal one-half of the compressive stress allowed. For ratios of compressive stress to shear between 0 and 1, proportionate shearing stresses shall be used. This gives a factor of safety of about 4.5.

The bonding stress between concrete and plain reinforcing bars may be assumed at 4 per cent of the compressive strength at 28 days, or 80 lb. per square inch for 2,000 lb. concrete; and in the case of drawn wire, 2 per cent, or 40 lb. per square inch for 2,000 lb. concrete. The factors of safety are about 4.5 and 2.25.

It is recommended that the modulus of elasticity of concrete in compression be assumed as $\frac{1}{15}$ of that of steel (2,000,000 lb. per square inch for a good 1:2:4 concrete 1 month old). While this assumption is not accurate, it will give safe results.

107. Rubble Concrete.—This is a concrete in which stones of a large size are handled and embedded separately. Rubble concrete construction is suitable only for massive work where the concrete is not less than 3 or 4 ft. thick. The saving over the cost of ordinary concrete is very little except in instances where the large stones can be very cheaply procured and handled. The usual procedure is to drop the large stones in the concrete and then spade the concrete around the stones so as to release the air and make a good bond. The large stones should be clean and the joints between the stones should be at least 4 in. thick and well filled with wet concrete. The concrete should be wet enough to flow readily around the stones.

G. CONCRETE STONE, BLOCK, AND BRICK

108. Definitions and Classifications.—Concrete stone may be defined as any precast concrete units of ordinary size which are used for construction purposes.

Concrete blocks are concrete stones which are considerably larger than ordinary brick. The blocks are of several shapes, varieties, and sizes. Most concrete blocks (except those used for purposes of ornamentation) are hollow so as to form air spaces in the walls and save weight and materials. There is no standard

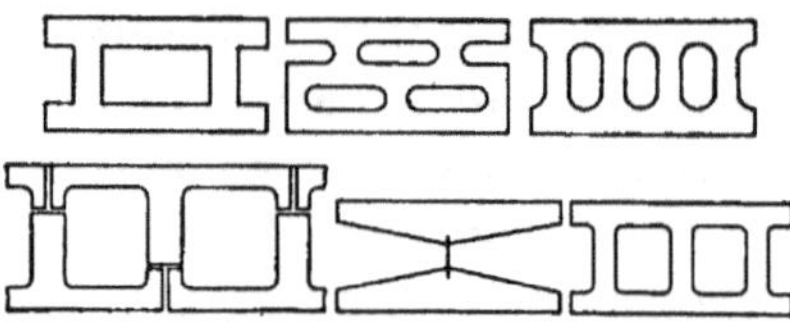

Fig. 21.—Cross sections of some concrete blocks.

size, though the length is usually 16 or 24 in., the height 8 or 9 in., and the thickness 8, 10, or 12 in.

Concrete brick usually are of the same size as ordinary building brick and are usually made solid, though some makes have grooves or hollows in the top and bottom.

Concrete stone may be divided into two classes according to use: (1) units for structural use primarily, such as ordinary solid or hollow blocks and brick; and (2) units designed primarily for purposes of architectural effect and ornamentation, such as the specially molded shapes or specially faced blocks or brick.

Concrete stone may also be classified according to the method of manufacture. The three general methods in use are the dry tamp, pressure, and wet-cast methods. Concrete stone for ornamental purposes is usually made in special molds by the dry tamp method.

109. Materials for Concrete Stone.—The materials should be chosen according to the principles governing the selection of materials for good concrete, except that the coarse aggregate should be a well-graded crushed stone or gravel that will pass a ¾-in. sieve and be retained on a ¼-in. sieve.

110. Proportions.—The proportions for concrete blocks should be 1 part of good portland cement to not over 2½ or 3 parts of good sand and to not over 3 or 4 parts of coarse aggregate. That is, the leanest allowable mix is a 1:3:4. When the coarse aggregate is omitted, the proportions should be 1 part of cement to not over 4 parts of good sand.

The limiting proportions for cement brick, which usually contain no coarse aggregate, are the same as those for concrete blocks.

In general, the proportions should be such that the concrete stone will pass the specifications given in a following article.

111. Consistency.—The best consistency is one where the mixture will just retain its shape when the molds are removed immediately after the concrete has been deposited and pressed in place. This consistency is much wetter than that usually used in the dry tamp method and a little wetter than that used in the pressure process. The consistency used in the wet-cast method is frequently too wet to secure the best results. A consistency that is too dry makes the concrete porous and increases its absorptive powers besides tending to reduce the strength.

112. Mixing and Molding.—The mixing may be done either by hand or machine. (See articles on hand and machine mixing for discussions of these methods.)

The molding may be done either by hand or machine except in the pressure process where a machine is required. Concrete blocks and brick for structural purposes are usually molded by machines while especially molded blocks are usually molded by hand. At present there are many different kinds of molding machines used in the manufacture of concrete stone. The construction of the machines varies with the consistency of the mix and the methods used for compacting the concrete.

The following are the three methods most frequently used in the manufacture of concrete blocks and brick:

(*a*) *Dry Tamp Process.*—The materials are first mixed to a damp consistency and are then thoroughly tamped in the molds by hand or machine tampers. Usually too little water is used in this process. This method is nearly always used in making concrete stone of special shape or surface finish as the molds may be made of any desired shape and size. The tamping is usually done by hand.

(*b*) *Pressure Process.*—A somewhat wetter mixture is used than in the dry tamp process. The concrete is then placed in the molds and pressure is applied either by mechanical levers or by a hydraulic piston.

(*c*) *Wet-cast Process.*—In this process the consistency is such that the concrete will readily flow. The mixture is poured into the molds and then thoroughly puddled to release any entrained air and to get the large particles away from the surfaces. No tamping or mechanical pressure is used. Frequently too much water is used.

In the first two methods the concrete is dry enough so that the molds can be removed immediately from the blocks, while in the last method the molds cannot be removed until after the concrete has set. If either of the first two methods is used, care should be taken to secure density and uniformity of compactness in the blocks.

113. Surface Finishes.—A variety of pleasing surface finishes may be secured with concrete blocks and brick. For a description of a number of ways of finishing the surface, see article on "Surface Finish."

Another way is to have one of the faces of the mold so shaped that the exposed face of the block, when placed in a wall, will have some pleasing shape such as some surface finish of stone masonry, etc. Special molds can be made to give a great variety of designs of cornices, rails, window seats, ornaments, etc.

Still another way is to place a facing layer of a selected fine material next to the face mold, this layer becoming intimately bonded with the body of the block in the process of molding.

Variety of color may be secured by using different colored stones or sands in the facing layer or by adding coloring matter when the materials are mixed. Only the purest mineral colors should be used, as coloring matter (especially when impure) tends to destroy the binding qualities of the cement. Sometimes the coloring may be secured by applying a cement stain to the desired surfaces.

114. Curing and Aging.—In curing, care should be taken to prevent the drying out of the blocks during their first hardening. After the molds are removed, the blocks should be protected from wind currents, sunlight, dry heat, and freezing for at least a week. During this time additional moisture should be supplied to the blocks by sprinkling, or some other method equally as good, at least once a day. After the first week the blocks should be sprinkled, or otherwise moistened, at occasional intervals until they are used. When cured by any natural process, concrete blocks should not be used for construction purposes until they are at least three weeks old.

The curing of concrete stone products may be accelerated by placing them (as soon as possible after they are removed from the molds) in an atmosphere of moist steam for at least 48 hours. The temperature of the curing room should be between 100 and 130 degrees Fahrenheit. The saturated steam provides heat and moisture and accelerates the hardening or setting of the concrete without causing the concrete to lose any of its moisture. After removal from the steam curing room, the concrete blocks should be stored for at least 8 days before using.

115. Properties of Concrete Blocks and Brick.—Concrete blocks and brick are usually not so strong as plain concretes of the same proportions. This is probably due to the fact that the consistencies used are not the ones which will give the greatest strength. Blocks made by the dry tamp and pressure processes often contain too little water while those made by the wet-cast process usually contain too much water.

Good concrete blocks and brick should pass the following tests:

(*a*) *Transverse Test.*—When subjected to transverse tests at an age of 28 days, the modulus of rupture should average more than 150 lb. per square inch and should not be less than 100 lb. per square inch in any individual case.

(*b*) *Compression Test.*—The ultimate compressive strength of solid blocks at the age of 28 days should average more than 1,500 lb. per square inch and should not be less than 1,000 lb. per square inch in any individual case.

The ultimate compressive strength of hollow blocks at the age of 28 days should average more than 1,000 lb. per square inch, and should not be less than 700 lb. per square inch in any individual case. In calculating the results, no reductions shall be made for the hollow spaces in the blocks.

The allowable working stress in compression should not exceed 107 lb. per square inch of gross area for hollow blocks, and 300 lb. per square inch of gross area for solid blocks.

(*c*) *Absorption Test.*—The samples shall be dried to constant weight at a temperature not exceeding 212 degrees Fahrenheit. After drying, the samples shall be immersed in clean water for 48 hours. The percentage of absorption (weight of water absorbed divided by the dry weight of the sample) should not average over 12 per cent and should not exceed 18 per cent in any individual case.

Full-sized blocks or brick shall be tested whenever possible.

The number of samples for any one test should not be less than three.

116. Uses of Concrete Blocks and Brick.—Blocks made of molded concrete can be used to advantage, as a substitute for solid concrete, brick, or stone, in the construction of walls that are thin or which carry only light loads, such as building walls, partitions, etc. Solid concrete is not satisfactory for such purposes on account of the expense of forms, the difficulty of securing a proper finish, and the prevention of the formation of cracks Concrete blocks are usually made of such shapes and sizes that they will form a wall containing hollow spaces, thus increasing the stability of the wall and forming dead air spaces as well as decreasing the weight.

Concrete brick are used as a substitute for ordinary building brick.

Especially molded and finished concrete blocks and brick are used for various purposes of architectural detail and ornamentation.

CHAPTER VI

BUILDING STONE

A. CLASSIFICATIONS AND DESCRIPTIONS

117. Building Stone in General.—Building stones include all of those stones or rocks that are used in masonry construction. The qualities which are the most important in stone used for construction purposes are cheapness, durability, strength, and beauty. In general, the hardest, densest, toughest, and most uniform stone will be the best stone to use.

The fitness of a stone for structural purposes may be approximately determined by the examination of a fresh fracture. This fracture should be bright, clean, sharp, without any loose grains and be free from a dull earthy appearance. An even fracture, when the surfaces of division are planes in definite positions, is characteristic of a crystalline structure. An uneven fracture, when the broken surface presents sharp projections, is characteristic of a granular structure. A slaty fracture gives an even surface for planes of division parallel to the laminations, and uneven surfaces for other directions of division. A conchoidal fracture presents smooth concave and convex surfaces, and is characteristic of a hard and compact structure. An earthy fracture leaves a rough, dull surface and indicates softness and brittleness.

The stone should contain no material, either in the form of seams or veins, that is not thoroughly cemented together. Stone containing much mica, pyrites, or glass seams usually are not very durable.

Only about half of all of the stone quarried is used for structural purposes, the other part being used for roads and pavements, crushed stone for railroad ballast and concrete, etc.

118. Classifications of Building Stone.—Building stone may be classified according to geological position, physical structure, or chemical composition. The geological position of a rock has but very little influence upon its properties as a building stone.

A. Geological Classification

1. *I*gneous rocks which are formed by a consolidation of the material from a fused or partly fused condition, such as greenstone, basalt, lava, etc.
2. Sedimentary rocks which are formed by a consolidation of material transported and deposited by water, such as sandstones, limestones, and clays.
3. Metamorphic rocks which are formed by a gradual change in the structure and character of igneous or sedimentary rocks due to their exposure to heat, water, pressure, etc. Some examples are the marbles and slates.

B. Physical Classification

1. Stratified rocks which are formed in layers, such as the sandstones, marbles, limestones, and some of the clays and slates. Their structure is either crystalline or granular or a combination of both.
2. Unstratified rocks which are not formed in layers. These rocks are usually made of crystalline grains strongly adhering together. Some examples are the granites, traps, basalts, etc.

C. Chemical Classification

1. Siliceous rocks in which silica is the most important chemical element, such as the granites, syenites, mica-slate, basalt, trap, quartz, sandstone, etc.
2. Argillaceous rocks (clayey rocks) in which the alumina governs the characteristic properties, such as the clays and slates. These stones are usually not very durable.
3. Calcareous rocks in which carbonate of lime is the important element, such as the marbles and limestones. The more compact of these stones are the more durable ones.

119. Granite, Gneiss, and Trap.—*Granite* is used more for structural purposes than any other igneous rock, and is the strongest and most durable of all the stones in common use. It is very hard and tough and, consequently, is difficult to cut and shape. However, it may be quarried in simple pieces without much difficulty as it breaks easily along its planes of weakness, which are at right angles to each other. Granite is used for foundations, base courses, walls, columns, steps, paving blocks, etc.

Gneiss has the same composition and about the same appearance as granite and is found in the same localities. It differs from granite by being usually arranged in more or less parallel layers, which makes the work of quarrying less difficult and expensive. This stone is used for foundation walls, courses, street paving, curbs, etc.

Trap is the strongest and one of the most durable of all building stones. It is also very tough and usually has no planes of cleavage. As it is difficult to quarry and work, trap is used very little for structural purposes.

120. Limestone, Marble, Sandstone, and Slate.—*Limestone* is a stone which contains calcium carbonate as its main constituent. It is very widely distributed and much used in building construction, probably ranking next to granite in this respect. Limestone differs greatly in color, composition, and structural qualities, because of the character of the deposits and their chemical composition.

Marble is a limestone which has been subjected to a metamorphic action and has had its structure changed to a more crystalline form. Its original color is usually changed and sometimes lost during this metamorphic action. Marble has a variety of colors, is very beautiful, and is much used for interior decorations. Often the name "marble" is improperly used by applying it to any limestone that will take a polish.

Sandstone is composed of grains of quartz sand which are cemented together by means of silica, iron oxide, calcium carbonate, or clayey materials to form a solid rock. This stone differs greatly in color, hardness, and durability, but some are very suitable for use in outside construction. The durability of the sandstone depends both upon its physical and its chemical composition. The best has silica as a cementing material and is usually soft when quarried but becomes harder upon exposure. Sandstone having iron oxide as a cementing material ranks next in durability, followed by that having calcium carbonate, while the one having clayey matter is the poorest. Sandstone is easier to quarry and work than limestone. Sandstone is used a great deal in building construction.

Slate is ordinarily composed of a siliceous clay which has been deposited in thin layers on a sea bed and later metamorphosed and compacted by pressure into a solid rock. Slate can be split into thin sheets and is tough, strong, and non-absorptive. It is used for roofing and some interior work in buildings.

B. STONE QUARRYING AND CUTTING

121. Hand Methods of Stone Quarrying.—No matter what method of quarrying is used, it is first necessary to remove the surface soil from the rock. The stone may be quarried by means of hand tools, machine tools, explosives, or by some combination of these methods.

Hand methods may be used when the stone occurs in thin beds. The principal tools used are the pick, crowbar, drill, hammer, wedge, and plug and feathers. In quarrying, rows of holes, which are from 3/8 to 3/4 in. in diameter and a few inches apart,are drilled in the rock by means of the drill and hammer. The distance between rows usually depends upon the dimensions of the desired stone. In drilling, a man holds the drill in one hand and drives it with a hammer in the other hand, rotating the drill a little between blows. Sometimes one man holds the drill and another man drives it with a heavy hammer or sledge. This kind of drill is called a jumper. Another kind of hand drill is the churn drill which is a heavy drill about 6 or 8 ft. long. This drill is raised by the workman who lets it fall in the desired place, then catches it on the rebound, rotates it a little while raising it, and lets it fall again, thus cutting a hole in the rock without the aid of a hammer. For deep holes, a churn drill is more economical than a jumper drill.

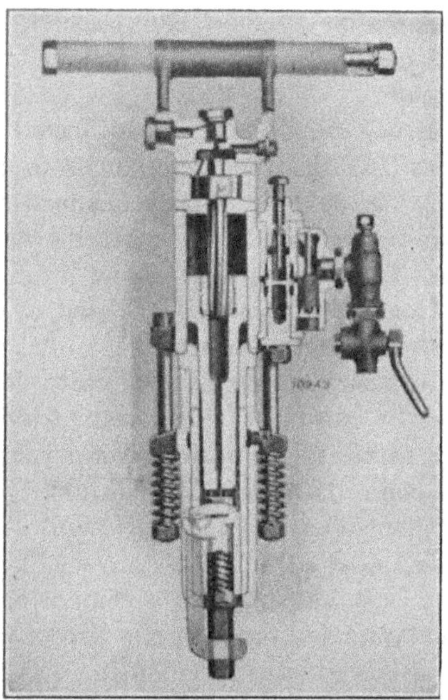

Fig. 22.—Cross section of "Jackhammer Sinker" drill. (*Ingersoll-Rand Co.*)

After the holes are drilled, a plug, inserted between two feathers, is placed in each hole. The plug is a narrow steel wedge with plane faces, and the feathers are wedges which are flat on one side and rounded on the other. Then the plugs in all of the holes are pounded in at the same time until they exert a force that is large enough to split the rock along the line of holes.

122. Machine Methods of Quarrying.—Machine methods include the use of machines driven by steam, compressed air, or electric motors to drill the holes or cut narrow channels in the rock. The machine drills are divided into two classes—percussion drills and rotary drills. In a percussion drill, the cutting tool resembles a hand drill. This drill is operated by a cylinder using steam or compressed air, or by an electric motor. An automatic device rotates the drill a little between strokes. In a rotary drill, the cutting tool is a hollow tube with a cutting edge made of sharp teeth or diamonds. This cutting edge is kept in contact with the rock while the drill is revolved, thus cutting a hole through the rock.

When large rectangular blocks of stones are desired, a special machine called a "channeler" is often used. This machine operates on a track or guide bars and carries a number of cutters which cut deep narrow channels in the rock as the machine slowly moves along.

After the holes are drilled, the rock is broken off by means of plugs and feathers or by means of explosives inserted in the holes.

Stone is rarely ever quarried by one method alone. The use of a combination of two methods is very common and that of three methods is not infrequent.

123. Explosives Used in Quarrying.—Explosives may be used instead of plugs and feathers to split off the stone after the drill holes have been made. The explosive is placed in the holes in the proper amounts, and the pressure (tamping) is provided by a little moist sand, clay, packed paper, etc. tamped on top. In the case of nitro-glycerine, a little water on the explosive provides all of the tamping necessary.

The explosives used in quarrying are usually gunpowder or dynamite and sometimes nitro-glycerine. The gunpowder must be a coarse, slow acting kind (commonly known as "blasting powder"). It is exploded by means of a fuse or electric spark.

Dynamite consists of some granular substance (such as sawdust) saturated with nitro-glycerine. True dynamite contains at least 50 per cent of nitro-glycerine, while the granular absorbent is an inert material. False dynamite may contain as little as 15 per cent of nitro-glycerine, but the absorbent material contains at least one other explosive. The other explosive is usually oxygen which is liberated in large quantities by the explosion and aids in effecting the complete combustion of the gases of the nitro-glycerine.

Nitro-glycerine is a fluid made by mixing glycerine with nitric or sulphuric acid. It is rarely used in quarrying as it acts too quickly and tends to shatter the rock very much.

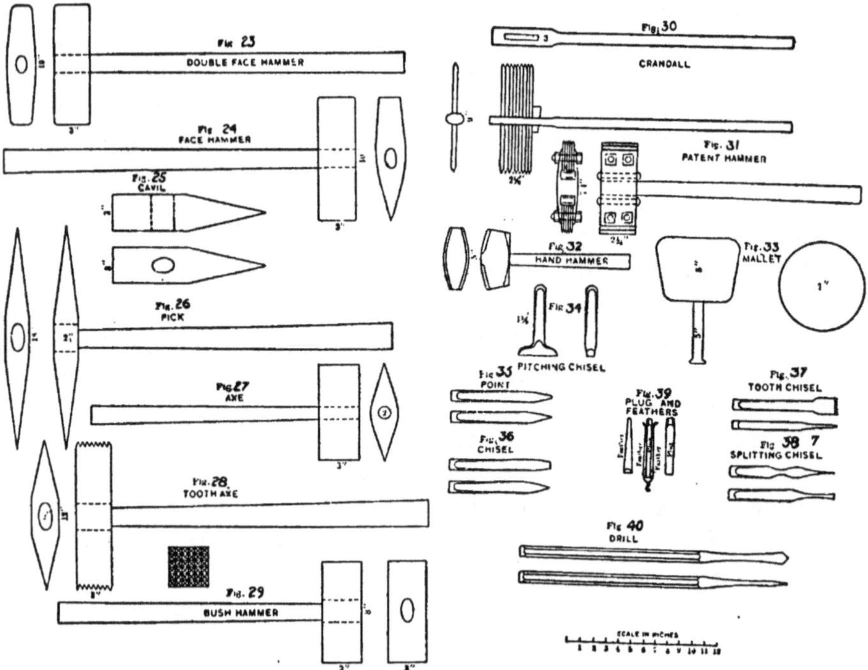

FIGS. 23–40.—Tools used in stone cutting.

Dynamite and nitro-glycerine are exploded by means of a percussion cap which is ignited by a fuse or an electric spark. A percussion cap is a hollow cylinder made of copper and has one end closed. This cylinder is about ¼ in. in diameter and 1 or 2 in. long. It contains a cement made of fulminate of mercury and some inert material.

124. Stone Cutting.—In stone cutting, various tools are used, such as stone hammers, picks, axes, points, chisels, mallets, pneumatic hammers, etc. These tools are often different in shape from ordinary tools of the same name. An examination of the sketches of these tools will furnish all of the description necessary, while their uses will be indicated in the following paragraphs.

Building stone are divided into three general classes and various subdivisions according to the finish of the surfaces (from *Trans. Am. Soc. Civ. Eng.*, Vol. 6).

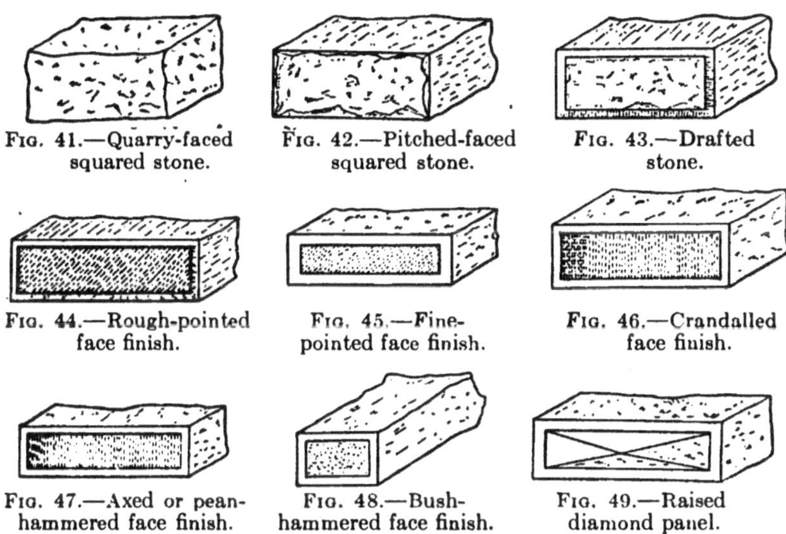

FIG. 41.—Quarry-faced squared stone.

FIG. 42.—Pitched-faced squared stone.

FIG. 43.—Drafted stone.

FIG. 44.—Rough-pointed face finish.

FIG. 45.—Fine-pointed face finish.

FIG. 46.—Crandalled face finish.

FIG. 47.—Axed or pean-hammered face finish.

FIG. 48.—Bush-hammered face finish.

FIG. 49.—Raised diamond panel.

1. Rough or Unsquared Stone

This class includes all stone that are used as they come from the quarry without any special preparation.

2. Stone Roughly Squared and Dressed

This class includes stone that are roughly dressed on beds and joints with the face hammer or axe. The distinction between this class and the third class lies in the closeness of the joints. When the dressing on the joints is such that the general thickness of mortar required is one-half inch or more, the stone properly belong to this class. There are three subdivisions:

(*a*) Quarry faced stone are those whose faces are left untouched as they come from the quarry.

(*b*) Pitched-faced stone are those in which the edges of the faces are made approximately true by the use of a pitching chisel.

(*c*) Drafted stone are those whose faces are surrounded by a chisel draft, the space inside being left rough. This method is not ordinarily used on stone of this (the second) class.

3. Cut Stone or Stone Accurately Squared and Finely Dressed

This class includes all stone dressed to smooth beds and joints so that the thickness of the mortar joints is less than one-half inch. As a rule, all of the edges of cut stone are drafted, and between the drafts the stone is smoothly dressed by one of the following methods. In massive construction work, the face of the stone is often left rough.

(a) *Rough-Pointed.*—The excess of material is removed by the pick or heavy point until the projections vary from ½ to 1 in. This method is much used on limestone and granite.

(b) *Fine-Pointed.*—The projections are less than ½ in. and the tool used is a fine point.

(c) *Crandalled.*—The same effect is produced as in fine pointed stone, except that the tool marks are more regular and with ⅛-in. projections. A stone is said to be cross crandalled when it is crandalled in both directions. The tool used is a crandall.

(d) *Axed or Paen Hammered.*—The face of the stone is covered with parallel chisel marks.

(e) *Tooth Axed.*—The same finish as fine-pointed stone, except that the tool used is a tooth axe.

(f) *Bush-Hammered.*—Where the roughnesses of the surface are pounded off with a bush hammer. This follows the rough pointing or tooth axing and is usually used only on limestone.

(g) *Rubbed.*—Where the sawn surfaces of the stone are smoothed by grit or sandstone. The method is used on marbles and sandstones.

(h) *Diamond Panels.*—Where the face inside the draft is cut to flat pyramidal forms.

C. PROPERTIES OF BUILDING STONE

125. Durability.—The durability of a stone depends upon its ability to resist the destructive actions due to the weather agencies. The determining factors of durability are the strue-ture, texture, and mineral composition of the stone. Imperfections, such as cracks, joint planes, etc., allow water to enter and disintegration to start through the action of frost. A coarse-grained or porous stone usually disintegrates more rapidly than does a fine-grained stone. Different mineral compounds in the stone also influence the durability. A stone containing silica or silicates is the most resistant to decay; followed by that containing aluminates; calcium and magnesium carbonates; iron compounds; and sulphides.

An increase in the durability of the stone may be secured by proper seasoning and surface finishing. When the stone is green, it contains much quarry sap and disintegrates much faster than after it is seasoned and the quarry sap has evaporated. In dressing the stone, care should be taken not to break up the grains too much and produce very small fissures through which the water may enter. A stone resists the effects of both pressure and weathering much better if it is placed on its natural bed.

The best way to determine the durability of the stone is to examine the surfaces of stone structures which have been exposed to atmospheric influences for years. There have been many artificial tests proposed, such as specific gravity, hardness or toughness, compression, cross-bending, shear, absorption, chemical, freezing, acid and microscopical tests, for determining the durability of the stone, but none of these tests give wholly satisfactory results.

The following table gives the estimated life of some of the building stones. The values are approximate, depending upon the variety of stone and place where it is used.

Approximate *Life* of Building Stone

Sandstone	may last from 20 to 200 years according to kind and place.
Limestone	may last from 20 to 40 years according to kind and place.
Marble	may last from 40 to 100 years according to kind and place.
Granite	may last from 75 to 200 years according to kind and place.
Gneiss	may last from 50 to 200 years according to kind and place.

126. Action of Frost, Wind, Rain, and Smoke.—Frost action, or freezing, disintegrates a stone only when the pores are practically filled with water before the freezing takes place. As a stone is not often used in such a way that the maximum amount of water is absorbed, it is very rare that a good building stone is injured by frost. Only a stone having a high absorptive power and a low structural strength is liable to be damaged by freezing.

A gentle wind has no effect on the stone but a heavy wind blows rain, dust, sand particles, etc. against the face of the rock. The sand and dust particles tend to wear away the surface by abrasion.

Rain, falling on the stone, penetrates the pores and tends to dissolve some of the chemicals in the stone. Rain water may contain some acids which are injurious to the stone. Also, there is the effect of pattering raindrops and flowing water wearing away the surface.

Smoke, which usually contains sulphuric or carbonic acid, has a bad effect on the stone, as these acids, as well as the nitric acid in the air, tend to cause disintegration.

127. Action of Fire.—Fires, such as destroy ordinary buildings, produce temperatures that are high enough to injure seriously the exposed building stone. The injury due to the combined action of both fire and water is usually much greater than that due to fire alone. Rapid cooling, by the application of water, of the exteriors of a highly heated stone tends to cause it to disintegrate.

Any stone expands upon being heated, but it only partly returns to its original dimensions on being cooled. This increase or set is very small, being about $3/100$ of 1 per cent of the length of the stone.

Granite usually cracks and spalls badly when exposed to fire. This stone has a low fire resistance.

Gneiss does not resist fire so well as granite does.

Limestone offers a high resistance to fire until a temperature of about 1,100 degrees Fahrenheit is reached and then the stone starts to decompose and crumble, due to the driving off of the carbon dioxide and the flaking of the quicklime formed. Limestone is injured more by slow cooling than by sudden cooling.

Marble cracks and spalls to some extent at temperatures below that at which calcination begins. After that temperature is reached, its action is like that of limestone.

Sandstone, especially if it is dense and non-porous, offers a better resistance to fire than other building stone. It cracks less than any other stone, and, if properly placed, these cracks will probably be horizontal. Sandstone having silica or lime carbonate for a cementing material resists fire better than a stone bound with iron oxide or clay.

128. Mechanical Properties of Building Stone.—The following tables give some average values of the strength and other mechanical properties of the principal building stone. It is to be noted that there are several varieties of each kind of stone and that the results obtained by testing any one variety may vary considerably from the average given in the table. However, most any variety of stone of good quality will give results equal to or greater than the average values given.

All of the values in these tables were obtained from the "American Civil Engineers' Pocket Book," except those in parenthesis which were obtained from other sources.

STRENGTH OF BUILDING STONE

Stone	Compression, pounds per square inch	Shear, pounds per square inch	Cross bending, pounds per square inch	Modulus of elasticity in compression, pounds per square inch
Granite	15,000(18,000)	2,000	1,500	7,000,000
Sandstone	8,000	1,500	1,200	3,000,000
Limestone	6,000(8,500)	1,000	1,200	7,000,000 (8,000,000)
Marble	10,000	1,400	1,400	8,000,000
Slate	15,000	.....	8,500	14,000,000

MECHANICAL PROPERTIES OF BUILDING STONE

Stone	Weight, pounds per cubic foot	Specific gravity	Per cent of absorption	Coefficient of expansion per degree Fahrenheit
Granite	170(165)	2.72(2.64)	0.5 (0.7)	0.0000040
Sandstone	150(145)	2.40(2.32)	3.0 (6.0)	0.0000055
Limestone	170(160)	2.72(2.56)	0.15(4.0)	0.0000045
Marble	170(165)	2.72(2.64)	0.10(0.4)	0.0000045
Slate	175	2.80	0.17(0.5)	0.0000058
Trap	185	2.96		

CHAPTER VII

BRICK AND OTHER CLAY PRODUCTS

A. CLASSIFICATIONS AND DEFINITIONS

129. Classifications.—Common brick may be classified according to use; as to position in the kiln when burned (see articles on kilns and burning); as to methods of manufacture; or as to form and shape. Brick and clay products are usually classified according to their uses.

Classification of Brick and Clay Products According to Their Uses.

1. *Building Brick.*—Used for ordinary building purposes.
   (*a*) Common building brick.
   (*b*) Face brick, pressed or re-pressed brick.
   (*c*) Enameled, glazed, and ornamental brick.
   (*d*) Hollow brick.
   (*e*) Sand lime brick.
   (*f*) Portland cement concrete blocks and brick.
2. *Paving Brick and Block.*—Vitrified brick or blocks used for paving.
3. *Fire Brick.*—Brick made so that they can withstand a high temperature.
4. *Terra Cotta.*—Used for structural purposes.
   (*a*) Architectural or decorative.
   (*b*) Blocks and lumber. Used for structural purposes.
   (*c*) Hollow building blocks and fireproofing material.
5. *Building Tile.*—Used for structural purposes.
   (*a*) Roofing tile.
   (*b*) Wall tile.
   (*c*) Floor tile.
6. *Drain Tile.*—Porous, non-vitrified, unglazed tile used for drainage purposes.
7. *Sewer Tile.*—Non-porous, vitrified, glazed tile used for sewerage purposes.

130. Definitions.—The following are definitions of some of the different kinds of brick. Other definitions will be found in the articles following.

Clay brick are made by molding, drying, and burning a proper mixture of sand and clay.

Sand lime brick are made from a mixture of sand and lime.

Terra cotta is made in about the same manner as ordinary clay brick, except that selected clays, that will burn to a desirable color with a slight natural glaze, and other materials are used.

Building tile are made in about the same manner and from about the same materials as pressed brick.

Tile and pipes are made by burning properly selected clay which has previously been molded in a suitable form.

Re-pressed brick are those made by pressing soft or stiff mud brick before firing.

Sewer brick are hard common brick or No. 2 paving brick used for sewers.

Face brick are usually pressed or re-pressed brick that are regular in shape and size and uniform in color. They are used for the outside of walls of buildings.

Feather edge brick have one edge thinner than the other. They are used in arches.

Compass brick have one edge shorter than the other. They are used for walls with curved surfaces.

Glazed or enameled brick have one face glazed or enameled.

B. MANUFACTURE OF CLAY BUILDING BRICK

131. The Clay.—Only the sedimentary clays are sufficiently homogeneous, fine, and plastic enough to be used in brick making. These clays consist principally of silicate of alumina together with a little lime, magnesia, and iron oxide. An excess of alumina makes the clay very plastic, but causes it to shrink, warp, and crack badly in drying, and also makes the clay very hard after burning. Uncombined silica, if not in excess (not over 25 per cent), tends to preserve the form of the brick, but an excess destroys the cohesion and makes the brick brittle and weak. Iron oxide makes the brick hard and strong. A little magnesia tends to decrease the shrinkage. Silicate of lime decreases the shrinkage, but it also softens the brick so that they will be distorted in burning. Carbonate of lime decomposes in burning and tends to cause the brick to disintegrate.

132. Hand Process of Making Brick.—The clay is first cleaned of all pebbles, dirt, etc., by washing, after which it is mixed with a moderate amount of water. To reduce the clay to a plastic mass, it is usually placed in a pug mill and then the proper amount of water is added. The pug mill consists, essentially, of a cylinder with revolving blades inside which cut up and mix the material. The clay may be made plastic by hand labor, but this process is very laborious. After "pugging," the plastic clay is pressed into the molds with the hands and tamped hard. The molds are sometimes sprinkled with water, but they are generally sprinkled with sand to keep the clay from sticking to them. This accounts for the names of "slop" and "sand" molding.

After molding, the brick are dried in air, often for several weeks, before they are fired or burned in a kiln. The length of time

FIG. 50.—Double shaft pug mill. (*American Clay Machinery Co.*)

required for drying depends on the weather conditions as well as on the composition of the brick.

If the brick are to be pressed, this must be done before they become too dry and hard. The press is a simple hand machine in which the brick are placed between plates or dies and then compressed by a piston operated by a hand lever.

133. Soft-mud Machine Process of Making Brick.—The three important ways of making machine-made brick are the soft-mud process, the stiff-mud process, and the pressed-brick process.

In the soft-mud process the clay is prepared as in the hand process and then reduced to a soft mud by the addition of water. The process of manufacture is practically the same as the hand process except that most of the work is done by machinery. The molding is done by a machine which presses the pugged clay into sanded molds by means of a plunger. Gang molds are used so that from 4 to 8 brick are molded at a time. Such a machine often has a capacity of from 8,000 to 12,000 brick per day.

The soft-mud brick are dried in the same way as the hand-made brick, except that sometimes a drying house is used to accelerate the drying.

134. Stiff-mud Machine Process of Making Brick.—In the stiff-mud process just enough water is added to the clay so

FIG. 51.—Soft mud type of brick machine with pug mill. (*American Clay Machinery Co.*)

that it will retain its shape after being molded under a moderate pressure. The consistency of this clay is like that of stiff mud, hence the name.

The molding is done by machines which are either of the auger or plunger types. In the auger type, the clay is forced through a die by means of an auger or screw working in a cylinder; while in the plunger type a simple piston or plunger is used instead of the auger. The die is an opening equal in size to the dimensions of an end or a side of the brick. When the clay comes through the die, it is forced out on a long table where it is cut in sections the size of an ordinary brick. If the cross-section of the bar of clay is the same as the end of a brick, the brick are called end cut; and if the section is the same as the side of a brick, they are called side cut.

Fig. 52.—Auger type of brick machine. (*American Clay Machinery Co.*)

Fig. 53.—Plunger type of brick machine. (*American Clay Machinery Co.*)

Often the brick are burned without any preliminary drying, but this is not good practice as they tend to crack and warp in the burning. It is better first to dry the brick for a time in air, or place them in a drying house where the drying is accelerated by means of steam pipes or hot air.

135. Pressed-brick Machine Process of Making Brick.—In the pressed-brick process, the clay is either used dry (containing less than 7 per cent of water) or semi-dry (containing more than 7 per cent of water but not so much water as the clay used in the stiff mud process). The clay is ground to the fineness of flour

Fig. 54.—Side cut brick table for plunger machine. (*American Clay Machinery Co.*)

before being delivered to the brick machine. This machine feeds the clay into the molds and then compresses it (under an enormous pressure) by means of plungers.

These brick require less drying than the other kinds, and, consequently, are often burned without any preliminary drying.

Pressed brick are very compact and strong, but they are thought to be less durable than those brick in which more water is used during the making.

136. Brick Kilns.—Brick kilns may be classified as intermittent or continuous kilns, according to their method of operation. The intermittent kilns are subdivided into updraught, downdraught, and up-and-down-draught kilns.

The old style updraught kiln is usually just a pile of brick about 20 or 30 ft. wide, 30 or 40 ft. long, and 10 or 15 ft. high. The brick are so piled as to form a number of arched openings extending through the pile. The sides of the pile are often plastered with mud and the top covered with dirt and sometimes roofed so as to keep in the heat. The more modern type of updraught kilns are built with permanent side walls. This updraught kiln is not so economical as the other types, as many

FIG. 55.—Dry clay brick press machine. (*American Clay Machinery Co.*)

FIG. 56.—Hundred and ten chamber Haigh continuous kiln. Five fires. (*American Clay Machinery Co.*)

(about half) of the brick have to be discarded on account of over-and-under burning.

A downdraught kiln has permanent walls, a floor, a tight roof, chimney, and furnaces. The floor has openings connecting with flues leading to the chimney. The heat is generated in outside ovens and enters the kiln in such a way that it reaches the top of the brick piles first and passes down through the piles to the openings in the floor, and then through the fines to the chimney. This kiln burns the brick very evenly and only a few of them have to be discarded on account of over-or under-burning.

The up-and-down-draught kiln is so arranged with two sets of furnaces that the heat from one furnace can be made to pass down through the brick while the heat from the other furnace can be made to pass up through the brick before passing to the chimney. This type of kiln burns the brick very uniformly.

Continuous kilns are of many types, but they all consist essentially of a series of chambers with flues between them and also between each chamber and the stack. When one chamber is fired, the heat can be made to traverse several other chambers before going to the chimney, thus preheating the brick in those chambers and securing an economy of fuel. Only one chamber needs to be out of operation at any one time because of the changing of the piles of brick.

137. Burning the Brick.—The fuel used usually depends upon local conditions. The furnaces in the kilns may be designed to use wood, coal, or gas. Wood is usually used in the old-style updraught kilns where the fires are built in the archways.

The brick are first subjected to a fire giving a moderate temperature until the moisture is expelled. Then the temperature is increased until the brick in the hottest part of the kiln are at a white heat, and the other bricks at a red heat. The fire is kept at this temperature until the burning of the brick is complete. Ordinary burning requires from 6 to 15 days.

When the burning is completed, the fires are stopped and all of the openings are closed so as to exclude any cool currents of air. Then the kiln and the brick are allowed to cool slowly for several days before the kiln is opened and the brick removed. This slow cooling "anneals" the brick and makes them tough.

The brick nearest the fire are usually overburned and are called arch or clinker brick. The brick in the coolest part of the kiln are usually underburned and these brick are called salmon or soft brick. All of the other brick in the kiln, which are properly burned, are called body, cherry, or hard brick and are the brick that are valuable for building purposes.

C. MANUFACTURE OF OTHER BRICK

138. Manufacture of Paving Brick.—These brick are used for paving purposes and should be hard, tough, and non-absorptive. Their manufacture differs from that of ordinary clay brick because they are burned at a much higher temperature (high enough to vitrify the brick) and also because the selection of a suitable clay is more limited.

Surface clays, impure fireclays, and shales have been used in the manufacture of paving brick, but the shales are the best and most used material at the present time. These clays occur in large bodies and are rocklike, but they are easily reduced to a powder. They are impure and have a range of vitrification often extending over 300 degrees Fahrenheit. The shale banks are usually worked with steam shovels.

When the clay arrives at the factory, it is crushed to a powder by grinding machinery and delivered to a pug mill where just enough water is mixed with the clay to make a stiff mud. The brick are molded as in the stiff-mud process and are repressed immediately after molding. This repressing makes the brick more uniform, rounds off the corners, and makes lugs on the sides so that the brick will be separated a little from each other when laid in a pavement. Sometimes the brick are wire cut and not repressed. This kind is called "wire cut lug brick." The brick are usually dried in a dry house before they are burned.

Paving brick are burned in a down-draught or a continuous kiln. The time required for burning is about ten days. The heat necessary is a bright cherry heat (from 1,500 to 2,000 degrees Fahrenheit) for shales, while only a red heat is reached in burning common-clay building brick. Different clays require different temperatures. After the brick are thoroughly burned, the kiln is tightly closed and allowed to cool slowly for several days. This tends to anneal the brick and make them more tough. Upon emptying the kiln the brick should be sorted into different classes. With shales, about 70 or 80 per cent of No. 1 paving brick are obtained.

139. Manufacture of Firebrick.—Firebrick may be classified as follows:

1. *Acid Brick.*—
   - (*a*) Fireclay brick.
   - (*b*) Silica brick.
   - (*c*) Ganister brick
2. *Basic Brick.*—
   - (*a*) Magnesia brick.
   - (*b*) Bauxite brick.

3. *Neutral Brick.*—

(*a*) Chromite brick.

Fireclay brick are made of ordinary fireclay mixed with a little flint clay, sand, burned fireclay, or other refractory material to prevent too great a shrinkage in burning and drying. The molding and drying may be done by any of the ordinary ways, while the firing is usually done in a down-draught or continuous kiln. A temperature of from 2,500 to 3,500 degrees Fahrenheit is required in the burning. The cooling should be fast until 2,500 degrees Fahrenheit is reached, and then it should be slow.

Silica firebrick are made of silica sand mixed with a little lime (about 2 per cent) to act as a binder. These firebrick are usually molded by hand, dried in a drying room, and fired in a down-draught kiln. The temperature required is from 2,600 to 3,200 degrees Fahrenheit. The cooling must be done very slowly and uniformly. A very good grade of silica firebrick can be made by the process used in making sand lime brick.

Ganister brick are made from ganister rock, which is a dense siliceous sandstone containing about 10 per cent of clay. The process of manufacture is the same as that for silica brick, except that no lime is added.

Magnesia brick are made from a mixture of caustic magnesia and sintered magnesia with a little iron oxide for a flux. The materials are ground and mixed; water is added in the pug mill; and the brick are molded under a heavy pressure. They must be carefully dried before being fired. The temperature required is from 3,300 to 3,450 degrees Fahrenheit. These brick warp and shrink badly.

Bauxite brick are made by mixing ground bauxite (containing more than 85 per cent of Al_2O_3) with about 25 per cent of clay in a pug mill; adding water; and molding by hand or with a stiff mud machine. The brick are burned at a temperature of about 2,800 degrees Fahrenheit. They are weak and shrink greatly in drying and burning.

Chromite brick are made from a mixture of chrome iron ore and fire clay or bauxite. The mixture contains about 50 per cent of chromium ore, 30 per cent of ferrous oxide, and 20 per cent of alumina and silica. The materials are ground, mixed, and molded under heavy pressure, as in the case of the silica brick. The burning temperature is about 3,000 degrees Fahrenheit.

140. Manufacture of Sand-lime Brick.—Sand-lime brick consist of a mass of sand cemented together with lime. There are several classes of these brick, but only one is of importance as a structural material. This class of brick is made of a mixture of sand and lime which is molded in a press and then subjected to steam under pressure.

The sand used should be well graded so as to have a low percentage of voids. The binding action between the sand and the lime is better with fine sand; hence, the sand should not be too coarse. A well-graded mixture has been found to have a low percentage of absorption. The sand should contain a sufficient proportion of fine quartz sand and it should also be clean and dry when mixed with the lime.

The lime used may be either a high-calcium or a dolomitic lime, but the former is preferable. The lime should be hydrated or slaked either before or at the time of mixing with the sand. The amount of lime varies from 5 to 10 per cent of the sand.

The thorough mixing of the sand, lime, and water is the most important part of the process of manufacture. It is preferable to mix the dry sand and the dry hydrated lime thoroughly and then add the water and mix again.

The mixture is molded into brick in the same manner as in the pressed-brick (dry clay) process. The pressure exerted by the machine on the mixture in the mold is about 15,000 lb. per square inch.

The brick are hardened by placing them in a closed hardening cylinder and subjecting them to steam under a pressure of about 125 lb. per square inch. The length of time required for hardening at this pressure is about 10 hours.

D. OTHER CLAY PRODUCTS

141. Terra Cotta.—Terra cotta is made in about the same way as ordinary clay brick, but it requires a carefully selected clay that will burn to a desirable color with a slight natural glaze. Usually, no single clay is used, but a mixture is made of several clays in order to obtain the desired effect.

Decorative terra cotta is usually made by hand molding and then dried and burned. Very great care must be taken to prevent distortion and discoloration during firing.

Terra cotta lumber and building blocks, which are used for

structural purposes, are usually made of a mixture of terra cotta clays and finely cut straw or sawdust. The method of manufacture is like that of the stiff-mud process, except that special care must be taken to prevent distortion and unequal heating in firing. The burning temperature is high enough to burn out all of the straw and sawdust and leave a light porous material. About all of the terra cotta lumber is hollow in construction with outside walls about 1 in. thick and partition walls about ¾ in. thick.

Terra cotta building blocks and fireproofing are the same as terra cotta lumber, except that no straw or sawdust is used and the firing temperature is high enough to vitrify the clay.

142. Building Tile.—Building tile may be divided according to use into roofing, wall, and floor tile.

Roofing tile are made in the same way as pressed brick, except that the flat forms may be made by the stiff-mud process. The clay is selected with greater care than in the case of ordinary brick. The shape of the tile may be flat, curved, or interlocking.

There are two kinds of wall tile called dust-pressed tile and plastic tile. The clay used in making dust-pressed wall tile may be a fireclay, shale clay, or a mixture of clays. The materials are ground, mixed, and then made up to the consistency of thin cream and strained through a silk screen. The water is drained off, the material dried, crushed to a powder, and slightly moistened by steam. The molding is done in a dry press and the tile are burned in fireclay boxes to keep the tile from coming in contact with the flames. After the first burning, a glaze may be applied and coloring matter added and the tile burned a second time to fuse the glaze. Plastic tile are made in the same way as dust-pressed tile, except that a mixture of soft clay and burned clay is used and the molding is done immediately after the mixing and the addition of the water. Most wall tile are dust-pressed tile.

An inferior product of building tile (and some kinds of terra cotta) is made from a finely divided clay similar to that used for ordinary clay brick. The stiff-mud process is used, and the tile are wire cut. The burning is the same as that given to ordinary clay building brick and is usually done in a down-draught kiln.

143. Drain Tile.—Drain tile are made from a red burning clay (shale clay), fireclay, surface clay, or from a mixture of clays like those used in making terra cotta lumber. The tile are made by the stiff mud process, and the burning is done at a temperature that aids in the production of a strong porous product that is not vitrified or glazed.

The tile may be classified according to the materials from which they are made, according to their use, or according to their

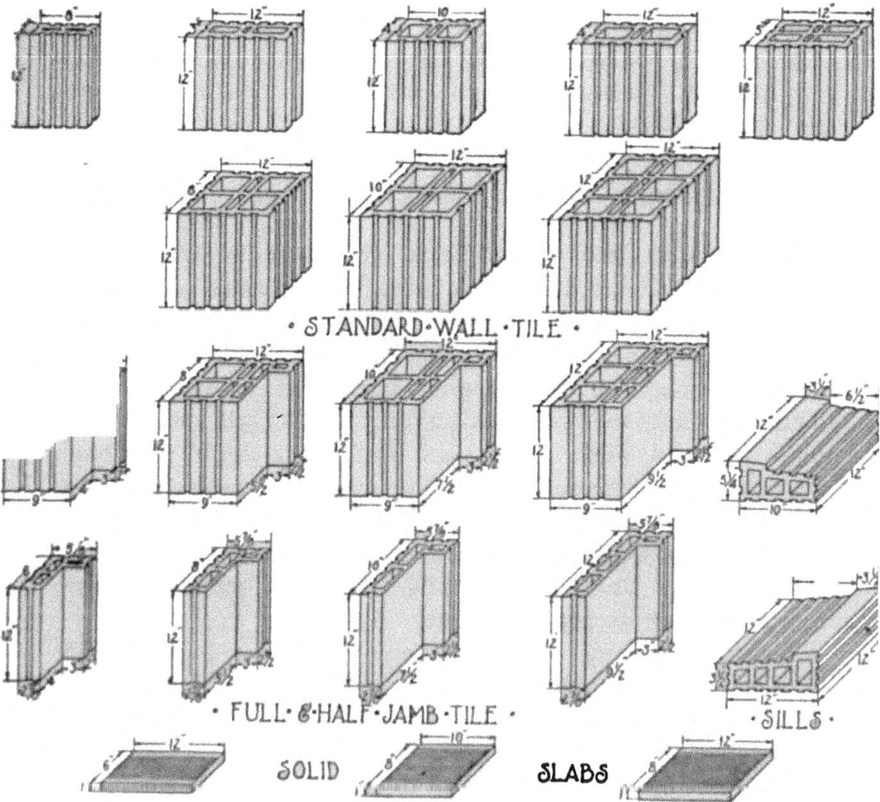

Fig. 57.—Standard wall and jamb tile.

general physical properties.

Drain tile are used for draining water from fields, roads, ditches, etc., and they must be porous so that the water can pass from the soil through the walls of the tile into the interior of the pipe.

144. Sewer Pipe.—Sewer pipe is made from such clays as will produce a non-porous tile with a low percentage of absorption.

The stiff-mud process of manufacture is used for ordinary pipes, and the dry-press process for those pipes having sockets at the end or which are of some special shape. The pipe is dried in a steam chamber and then burned in a down-draught kiln. When the temperature reaches about 2,100 degrees Fahrenheit, common salt is thrown on the kiln fires. The sodium vapors from the salt combine with the clay and form a hard glaze on the surface of the pipe and thus make the pipe very non-absorptive. Sewer pipe may be classified in the same ways as drain tile.

As sewer pipes are used for carrying sewage, it is important that they be non-porous and non-absorptive and that they have good tight joints.

E. PROPERTIES OF BRICK AND OTHER CLAY PRODUCTS

145. General Properties of Brick. *Requisites of Good Brick.*— A good brick should have plane faces, parallel sides, sharp edges and angles, a fine, compact, uniform texture, and it should contain no cracks, fissures, air bubbles, pebbles, lumps of lime, etc. A brick should give a clear ringing sound when struck with the hammer or another brick. A paving brick should be hard and tough to resist wear and impact, and it should be free from laminations or seams so that it will wear uniformly in a pavement.

Common clay building brick of good quality weigh about 125 lb. per cubic foot; face or pressed brick about 135 lb. per cubic foot; sand-lime brick about 115 lb. per cubic foot; and paving brick about 150 lb. per cubic foot.

The sizes of brick vary in different countries and different localities. The standard size for common brick in America is 8¼ by 4 by 2¼ in., and for paving brick 8½ by 4 by 2½ in. Paving blocks are about 3 by 4 by 9 in. in size.

146. Absorption of Brick and Building Tile.—Formerly, it was thought that if a brick would absorb much water it was not so durable as other brick and was more liable to destruction by frost, but this opinion has not been substantiated by tests. The absorptive power of a brick depends somewhat on its compactness but more on the chemical composition of the clay. There appears to be no close relation between the absorptive power and the strength and durability of the brick. The following table shows the approximate range of absorption in different kinds of brick which have been immersed in water for 48 hours.

Absorption of Brick and Building Tile

Percentages are based on the weight of the dry brick

Kind of Brick	Percentage of Absorption
Common clay building brick	12 to 18
Pressed or face brick	6 to 12
Sand-lime brick	12 to 15
Paving brick and blocks	1 to 3
Fireclay brick	8 to 12
Unglazed terra cotta blocks and building tiles	10 to 15

147. Compressive Strength of Brick and Building Tile.—The compression test of brick is only of relative value for comparing different kinds of brick, because, when a brick is used in masonry, its crushing strength is not of much importance unless the mortar used with the brick has nearly the same strength. Ordinary mortar used in brick masonry is generally very much weaker than the brick. Soaking a brick in water tends to decrease its strength in compression. The following table will give an idea of the average compressive strength of good brick:

Compressive Strength of Brick and Building Tile

Kind of Brick	Strength in Pounds Per Square Inch
Average good clay building brick	4,000
Pressed brick	8,000
Sand-lime brick	3,000 to 4,000
Paving brick and blocks	10,000
Fireclay brick	3,000 to 6,000
Terra cotta blocks and building tile	4,000
Architectural terra cotta	3,000

148. Transverse Strength of Brick and Building Tile.—The transverse tests are easy to make and they give results that are definite and which furnish the best indications of the quality of the brick. Transverse tests afford an indication of the toughness and also of the ability of the brick to resist ordinary failures in brick walls. In masonry walls the mortar usually fails first and squeezes out, thus setting up bending stresses in the brick which cause them to fail. The appearance of the fractured surface is a good indication of the care with which the brick have been made. The following table gives an average range of values:

CROSS-BENDING STRENGTH OF BRICK AND BUILDING TILE

Kind of Brick	Modulus of Rupture in Pounds per Square Inch
Common clay building brick	500 to 1,000
Pressed or face brick	600 to 1,200
Sand-lime brick	300 to 600
Paving brick and blocks	1,500 to 2,500
Fireclay brick	300 to 600
Unglazed terra cotta blocks and building tile	500 to 1,000

149. Shearing Strength of Brick and Building Tile.—The shearing strength of brick is of but little importance and the tests are very hard to make properly. Tests made at the Watertown Arsenal gave the following values. Results for terra cotta and building tile were taken from another source.

SHEARING STRENGTH OF BRICK AND BUILDING TILE

Kind of Brick	Strength in Pounds per Square Inch
Common clay building brick	1,000 to 1,500
Pressed or face brick	800 to 1,200
Sand-lime brick	500 to 1,000
Paving brick and blocks	1,200 to 1,800
Fireclay brick	500 to 1,000
Unglazed terra cotta blocks and building tile	600 to 1,200

150. Modulus of Elasticity of Brick and Building Tile.—The modulus of elasticity of brick in compression is not a constant quantity because the stress strain curve in compression is a curved line throughout its length, similar to the stress strain curves for concretes and mortars. The following are average values of the modulus of elasticity in compression for loads not exceeding one-fourth of the ultimate strength.

MODULUS OF ELASTICITY IN COMPRESSION OF BRICK AND BUILDING TILE

Kind of Brick	Modulus of Elasticity in Pounds per Square Inch
Common clay building brick	1,500,000 to 2,500,000
Pressed or face brick	2,000,000 to 3,000,000
Sand-lime brick	800,000 to 1,200,000
Paving brick and blocks	4,000,000 to 8,000,000
Unglazed terra cotta blocks and building tile	1,500,000 to 3,000,000

151. Properties of Drain Tile. *Requisites.*—All drain tile should be free from visible grains of caustic lime, iron pyrites, or other minerals which are known to cause slaking or the disintegrating of the tile. The drain tile should be of the proper shape, diameter, and length; uniform in structure, smooth on the inside, free from cracks and checks that would appreciably lower the strength; properly burned; and should give a clear ring when stood on end and tapped with a light hammer.

Drain tile are divided into three classes (farm drain tile, standard drain tile, and extra quality drain tile) according to quality and use. The physical tests include strength and absorption tests and sometimes freezing tests. Any good drain tile should easily pass the following minimum requirements for strength and not absorb more water than the maximum values given below:

MINIMUM REQUIREMENTS FOR STRENGTH OF DRAIN TILE

Internal diameter of pipe, inches	Average supporting strength in pounds per lineal foot		
	Farm drain tile	Standard drain tile	Extra quality drain tile
4	800	1,200	1,600
8	800	1,200	1,600
12	800	1,200	1,600
16	1,000	1,300	1,700
20	.....	1,500	2,000
24	.....	1,700	2,400
30	.....	2,000	3,000
36	.....	2,300	3,600
42	.....	2,600	4,200

MAXIMUM ALLOWABLE ABSORPTION FOR DRAIN TILE

Standard boiling test. All values are percentages of the dry weight.

Materials used in making	Farm drain tile, per cent	Standard drain tile, per cent	Extra quality drain tile, per cent
Shale and fireclay tile......	11	9	7
Surface clay tile...........	14	13	11
Concrete tile..............	12	11	10

152. Properties of Sewer Pipes. *Requisites.*—Sewer pipes should be of the hub and spigot type preferably, properly made and burned, vitrified, and salt glazed. All pipes should be of proper dimensions, straight, sound, well glazed throughout, smooth on the inside, free from blisters, lumps, or flakes which are broken or are larger than allowed by the specifications, and free from fire checks and cracks extending through the thickness of the pipe. The thickness of the walls should be at least 1/12 of the inside diameter.

The thickness of the walls of ordinary sewer pipes is less than 1/12 of the inside diameter, while that of "double strength" pipes is equal to 1/12 of the inside diameter.

The following table gives the minimum strength requirements of the Tentative (Proposed) Specifications for Sewer Pipes of the American Society for Testing Materials, the Specifications of the City of Brooklyn for Sewer Pipes and the results of strength tests made in that city. These results are higher than would ordinarily be expected from tests on ordinary salt-glazed and vitrified-clay sewer pipes. Any good sewer pipe should be able to pass the A. S. T. M. specifications.

The requirements for the specification for absorption have not yet been decided upon. However, good salt-glazed vitrified-clay sewer pipes should not absorb more than 3 per cent of water on the average or more than 5 per cent in any individual case.

MINIMUM STRENGTH REQUIREMENTS AND RESULTS OF TESTS OF SEWER PIPES

Average supporting strengths in pounds per lineal foot.

Inside diameter, inches	A. S. T. M. tentative specifications	Brooklyn specifications	Brooklyn test results
6	1,430	1,000	4,275
8	1,430		
9	.....	1,050	3,983
10	1,570		
12	1,710	1,150	4,696
15	1,960	1,300	5,046
18	2,200	1,450	6,311
21	2,590		
24	3,070	2,000	9,866
30	3,690		
36	4,400		
42	5,030		

CHAPTER VIII

STONE AND BRICK MASONRY

A. STONE MASONRY

153. Stone Masonry in General.—Stone masonry includes all masonry in which stone form the most important part. When mortar is used with stone masonry, it is called "wet" or "mortar" masonry. When no mortar is used, it is called "dry" masonry.

Stone masonry is one of the oldest forms of construction known to mankind. It has been used by practically all peoples throughout all ages and in all countries where the stones could be readily obtained.

Structures made of stone masonry are very durable and some of them have been in use for more than a hundred years.

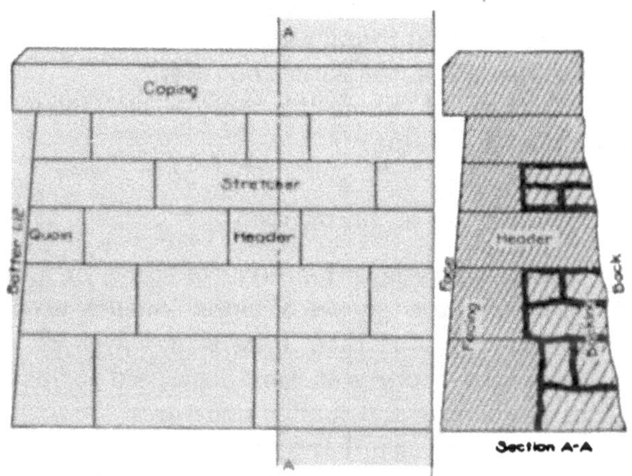

FIG. 58.—Range masonry. Sketch showing arrangement and names of parts.

Stone masonry is used for various buildings, retaining walls, dams, piers, abutments, arches, bridges, paving, culverts, foundations, etc.

154. Definitions.—The following are definitions of some of the terms often used in connection with stone (and brick) masonry. Other definitions will be found in the articles following.

Batter is the slope of the surface of the wall.

Coping is a course of heavy stones laid on top of a wall to protect it.

Course is a horizontal layer of stones in a wall.

Cramps are bars of iron or steel having their ends bent at right angles to the body of the bar. These ends enter holes in adjacent stones to keep the stones from separating.

Dowels are short straight bars of iron or steel which enter holes in adjacent stones which are above each other to prevent one stone from slipping on the other.

Face is the front surface of the wall.

Back is the rear surface of the wall.

Facing is the stone or other material which forms the face of the wall.

Filling is the material in the interior of the wall.

Backing is the material forming the back of the wall.

Quoin is a stone laid in the corner of a wall.

155. Classification of Stone Masonry.—The following is a classification of stone masonry:

A. Dry Masonry

1. Slope wall masonry is a thin layer of stone, or an inclined wall of stone, built against slopes of embankments, excavations, river banks, etc., to protect them from rain, waves, or weather.
2. Stone paving is a dry stone masonry used for paving the floors or ends of culverts and similar structures.
3. Riprap is stone of any shape or size placed on river banks or around piers, abutments, etc., to prevent wash and scour by the water. The stone may be dumped in place, but they are more effective if arranged by hand.

B. Wet or Mortar Masonry

1. Rubble masonry which is composed of rough unsquared stone.
   - (*a*) Coursed rubble in which the stone are leveled off at specified heights to an approximately level surface.
   - (*b*) Uncoursed rubble which is laid without any attempt at regular courses.
2. Squared stone masonry is masonry in which the stone are roughly squared and roughly dressed on beds and joints. The

thickness of the mortar required in the joints is more than ½ in. This class may be subdivided according to the facing of the stone into:

(*a*) Pitched faced masonry.

(*b*) Quarry faced masonry.

Or this class may be divided according to the manner in which the stone are laid:

(*a*) Range work which is laid in courses.

(*b*) Broken range which is laid in broken courses.

(*c*) Random masonry which is laid with no attempt at courses.

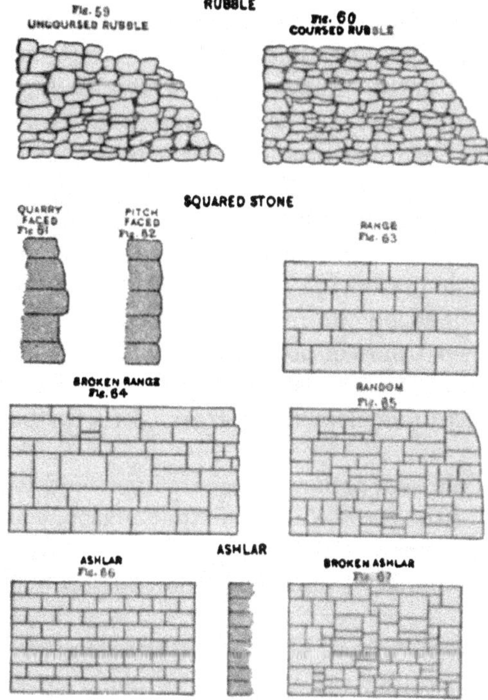

Figs. 59–67.—Stone masonry.

3. Cut stone or ashlar masonry which is composed of any of the kinds of cut stone where the required thickness of the mortar joints is less than ½ in.

This class is usually subdivided as follows:

(*a*) Coursed ashlar in which the courses are continuous (range work).

(*b*) Broken ashlar in which the courses are broken and not continuous (broken range).

156. Mortar for Stone Masonry.—Mortar for stone masonry has three functions: (1) to form a bed or cushion for the stone so as to distribute the pressure uniformly; (2) to bind the wall together into a solid whole; and (3) to fill the spaces and voids in the masonry and keep out the water. Also, a good mortar should be soft and plastic so that it will work properly besides being capable of hardening and becoming strong, dense, and impervious.

In general, the kind of mortar used depends upon the kind of masonry and the loads the masonry is to bear.

Lime mortar is usually used with rubble masonry, often with squared stone masonry, and rarely with cut stone or ashlar masonry. Probably more lime mortar is used than any other kind as it is very suitable for masonry where the loads are small. This mortar is composed of 1 part of lime paste to 2½ to 3 parts of good, clean, fine, sharp sand. See chapter on "Limes and Lime Mortars" for a further discussion of this mortar.

A Portland cement mortar is usually used with cut stone or ashlar masonry, sometimes with squared stone masonry, and rarely with rubble masonry. This mortar should always be used where the unit load on the masonry is large. The proportions usually vary from 1 part of portland cement to from 1 to 4 parts of sand according to the strength desired. See chapter on "Portland Cement and Cement Mortars" for a further discussion of this mortar.

A mortar made of Portland cement, lime, and sand may be used with any of the three classes of stone masonry as conditions permit. This mortar is stronger than lime mortar, and weaker, more plastic, and more impervious than an ordinary portland cement mortar.

157. Dressing of Stone Masonry.—Dressing is the cutting of the side and bed joints of the stone to plane surfaces, usually at right angles to each other. Care should be taken to make the bed a plane surface so that the pressure will be distributed evenly over all of the stone and also so that the bending stresses in the stone will be a minimum. Great smoothness is not desirable in the joints as slightly rough surfaces offer a greater resistance to slipping and also tend to increase the adhesion of the mortar.

158. Bond in Stone Masonry.—Bond in masonry is the overlapping of the stone so as to tie the wall together both longitudinally and transversely, and is of great importance to the strength of the wall. The stone in any course should be laid so that they will overlap or break joints with those in the course below, and in such a manner that each stone will be supported by two (or three) below and will aid in supporting at least two above it in the wall.

A very strong bond is made by laying some of the stone with their greatest dimension perpendicular to the face of the wall. Such stone are called "headers." In thin walls the headers should be long enough to extend clear through the wall.

Stone laid with their greatest dimension parallel with the face of the wall are called "stretchers."

159. Backing in Stone Masonry.—Ashlar or cut-stone masonry is usually backed with coursed rubble masonry and sometimes

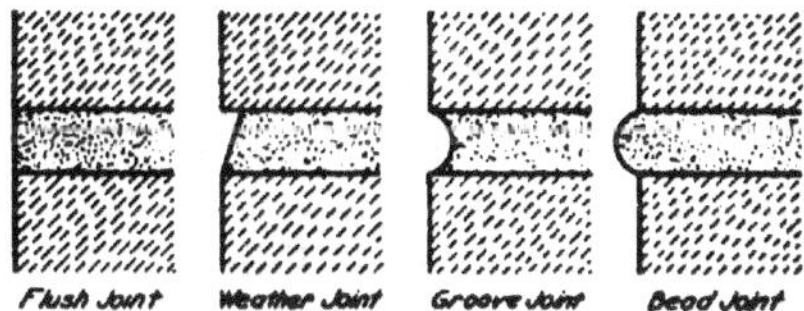

FIG. 68.—Methods of finishing horizontal joints.

with brick masonry. Squared-stone masonry is sometimes backed with rubble or brick masonry.

Great care should be taken to secure a good bond between the facing masonry and the backing. Headers should be frequently used. For the best bond, the backing should be built up with the facing masonry.

160. Pointing of Stone Masonry.—Pointing is the refilling of the edges of the joints in the masonry as compactly as possible and to a depth of about 1 in. with mortar especially prepared for that purpose. Only a very good cement mortar such as a 1:1 or a 1:2 portland cement mortar should be used. Sometimes a good lime mortar is used for some classes of masonry.

The four general ways of pointing the edges of the horizontal joints in cut stone masonry are flush joints, weather joints, grooved joints, and bead joints. The vertical joints are pointed in the same manner as the horizontal ones, except that in weather joints, the vertical joints are made flush. Care should be taken not to reverse the slope in the weather joint as this would allow water to collect in the joint and tend to weaken the masonry.

161. General Rules for Laying Stone Masonry.—The following general rules for laying stone masonry are taken from "Baker's Masonry Construction." These principles apply to all classes of stone masonry.

1. The largest stone should be used in the foundation to give the greatest strength and lessen the danger of unequal settlement.

2. A stone should be laid upon its broadest face, since then there is better opportunity to fill the spaces between the stones.

3. For the sake of appearance, the larger stone should be placed in the lower courses, the thickness of the courses decreasing gradually toward the top of the wall.

4. Stratified stone should be laid upon their natural bed, that is, with the strata perpendicular to the pressure, since they are then stronger and more durable.

5. The masonry should be built in courses perpendicular to the pressure it is to bear.

6. To bind the wall together laterally a stone in any course should break joints with or overlap the stone in the course below; that is, the joints parallel to the pressure in two adjoining courses should not be too nearly in the same line. This is briefly stated by saying that the wall shall have sufficient lateral bond.

7. To bind the wall together transversely there should be a considerable number of headers extending from the front to the back of thin walls or from the outside to the interior of thick walls; that is, the wall should have sufficient transverse bond.

8. The surface of all porous stone should be moistened before being bedded, to prevent the stone from absorbing the moisture from the mortar and thereby causing it to become a friable mass.

9. The spaces between the back ends of the adjoining stone should be as small as possible, and these spaces and the joints between the stone should be filled with mortar.

10. If it is necessary to move a stone after it has been placed upon the mortar bed, it should be lifted clear and reset, as attempting to slide it tends to loosen stones already laid and destroy the adhesion, and thereby injure the strength of the wall.

11. An unseasoned stone should not be laid in the wall if there is any likelihood of its being frozen before it has seasoned.

162. Waterproofing Stone Masonry.—As most of the stone used for masonry is practically impermeable, the weak or permeable part of the masonry is the joints. If good mortar is used and all of the spaces between the stone are filled, practically no water will pass through. Care should be taken to see that all joints are carefully pointed and that all cracks are filled with good mortar.

Washing or painting the surface of the masonry exposed to the water with a waterproofing compound (such as a soap and alum solution, hot tar, asphalt, etc.) will aid in making the masonry water-tight.

Sometimes the masonry is made more water-tight by incorporating a layer of felt or tar paper, painted on both sides with tar or asphalt, in the wall. Care should be taken to make the ends of the felt or paper overlap so that no cracks or holes extend through this waterproofing layer. Such a waterproofing layer is sometimes applied to the face or back of the wall instead of being built in the wall.

163. Cleaning Stone Masonry.—When the masonry is completed, the surfaces should be cleaned to remove any dirt, mortar, etc. adhering to the wall. The cleaning is usually done by brushing with stiff brushes and then washing with water.

Frequently, the stone work in buildings or other structures becomes soiled by dirt in the air, or smoke. The stone work may be cleaned with soap and water or by brushing and then washing with soap and water. Sometimes washing with a dilute acid solution aids in cleaning and brightening the surface. The use of the sand blast is very effective for cleaning purposes.

164. Strength and Other Properties of Stone Masonry.—The strength of stone masonry depends not only upon the strength of the stone in compression but also upon the accuracy of the dressing, the bond between the stones, and the thickness and strength of the mortar. In practically all of the observed failures of stone masonry under compression, the mortar failed first and squeezed out, thus causing bending stresses in the stone which resulted in their failure by tension in cross bending. About the only practical way of determining the strength of good stone masonry is to note the loads that have been carried by it without failure. There are several structures of first-class masonry carrying loads of approximately 400 lb. per square inch without showing any signs of failure.

Ashlar or cut stone masonry is the best of all stone masonry in quality, and it is used in all important structures where strength and stability are required. The stone used should not be longer than 3 to 5 times their depth, nor wider than 2 to 3 times the depth, depending to some extent upon the strength of the stone.

The weight per cubic foot of stone masonry may be taken at about 5 lb. less than that of the stone used.

The modulus of elasticity of stone masonry in compression is about 2,000,000 lb. per square inch for rubble masonry and about 4,000,000 lb. per square inch for ashlar masonry. These values are approximate only and depend to a large extent upon the stone and mortar used as well as the class of masonry and the care with which it is constructed.

The amount of mortar required for ashlar masonry is about 2 or 2½ cu. ft. per cubic yard of masonry; for squared stone masonry, from 3½ to 5 cu. ft. per cubic yard of masonry; and for rubble masonry, from 7½ to 10 cu. ft. per cubic yard of masonry.

The coefficient of expansion of stone masonry is about 0.0000035 per degree Fahrenheit.

The tensile strength of masonry is very small, and, therefore, stone masonry should not be designed to carry any tension. As stone masonry is very weak in tension, it is also weak in cross bending, and should not be expected to carry transverse loads.

165. Safe Loads for Stone Masonry.—Safe loads for stone masonry in tension and cross bending should be considered as zero, except in special cases where there are special designs and constructions.

The working stress in shear should be taken at one-fourth of the safe working stress in compression. See tables following for safe working stress in compression.

From an examination of the loads carried by different classes of the *best* stone masonry without failure, the values in the following table may be assumed, provided that each kind of stone masonry is the *best* of its class:

Safe Loads in Compression for the Best Stone Masonry

Kind of Masonry	Good Ordinary Mortar, Pounds Per Square Inch	Portland Cement Mortar 1:2 Mix, Pounds Per Square Inch
Rubble	140 to 200	
Squared stone	200 to 280	
Limestone ashlar	280 to 350	600
Sandstone ashlar	250 to 320	500
Granite ashlar	350 to 400	700

The building laws (1907) of the city of Chicago gave the following allowable safe loads in compression for masonry.

These values were recommended to the city by a large committee composed of the leading architects and engineers of Chicago.

City of Chicago

Allowable Unit Stresses for Masonry in Compression

All values are in pounds per square inch

Kind of Masonry	Lime Mortar	Portland Cement Mortar
Rubble masonry, uncoursed	60	100
Rubble masonry, coursed	120	200
Ashlar masonry, coursed limestone	...	400
Ashlar masonry, coursed sandstone	...	400
Ashlar masonry, coursed granite	...	600

B. BRICK AND HOLLOW TILE MASONRY

166. Brick Masonry in General.—Brick masonry includes all forms of masonry composed of brick and mortar, such as walls of buildings, backing for stone and concrete masonry, sewers, tunnels, facing for stone or concrete masonry, arches, etc., and sometimes culverts, piers, abutments, and bridges.

At the present time, brick masonry is much used as a building material. Good brick masonry is the equal of stone masonry in strength, durability, and appearance. Some of the advantages of brick masonry are: brick resist the action of fire, weather, and the acids in the atmosphere; brick can be secured in any locality, and of most any size, shape, and color; brick are easy to lay in a wall; brick masonry is as durable as stone masonry; brick masonry is as strong as stone masonry, with the exception of cut stone masonry; brick masonry is often cheaper than stone masonry. Some of the disadvantages of brick masonry are: it requires skill to lay a good wall; poor mortar is often used in the laying of the brick, thus making a wall that is neither strong nor durable.

167. Mortar for Brick Masonry.—As in stone masonry, the mortar in brick masonry has three functions to perform: namely, (1) to form a bed or cushion to take up any inequalities in the brick and to distribute the pressure uniformly; (2) to bind the wall into a solid mass; and (3) to fill the spaces and voids between the brick in the masonry and keep out the water.

In general, the mortar should be chosen to suit the character of the masonry and the loads that it is to bear. For strong and impervious masonry or masonry which may be under water, a Portland cement mortar (of a mix varying from a 1:2 to 1:4

according to conditions) should be used. For small loads a good lime mortar is suitable, while for medium loads, a mortar composed of Portland cement, lime, and sand may be used. Clay or loam should never be used in place of the sand. At the present time most of the brick masonry is laid in lime mortar on account of its cheapness.

The quantity of mortar required for brick masonry depends upon the size of the brick and the thickness of the joints. As many of the building brick are about of the same size, most of the variation in the quantity of mortar needed is due to the thickness of the joints.

The thickness of joints in brick masonry may vary from 1/8 to 3/4 in., depending on the kind of brick used and the architectural effect desired. For pressed brick, a joint of about 3/16 in. is desirable, while for ordinary brick the joint should be from 1/4 to 3/8 in. thick for good work. For ordinary backing and filling and rough work, the joints are usually from 3/8 to 1/2 in. thick.

168. Laying the Brick.—The following principles apply to all brick laying:

1. All brick should be thoroughly wet before laying, so that they will not absorb the water from the mortar. While this wetting is important, it is often neglected.
2. The brick should be laid in a truly horizontal position except in special cases.
3. The top edge of a brick should be laid to a stretched string.
4. The masonry should be built in courses perpendicular to the pressure it is to bear.
5. Each course should break joints with the courses immediately above and below it. There should be sufficient longitudinal bond.
6. Sufficient transverse bond should be provided.
7. The spaces between the brick should be completely filled with mortar.
8. In laying the brick, a layer of mortar should first be spread over the last course of brick.
9. The brick should be firmly pressed in place in this mortar with a sliding motion which will force the mortar to fill the joint.
10. The excess mortar squeezed out on the face of the wall should be removed with the trowel and applied to the end of the brick so as to aid in filling the next joint.

169. Improvements in Brick Laying.—During the last few years, the laying of brick has been greatly expedited by the use of three innovations: the packet, the special scaffold, and the fountain trowel. By the use of these innovations, together with proper instructions, a skilled bricklayer can lay three or four times as many brick as he could before.

The packet is a small wooden frame or tray upon which two rows of ten brick each are placed on edge in such a position that the mason can put his fingers under each brick while it is upon the packet. The brick are placed on the packets when they are unloaded from the wagon or car, and are transported on the packets to the scaffold. The brick may be sorted as they are placed on the packets.

The special scaffold is simply a shelf or bench about two and a half feet above the platform on which the mason stands. The packets are placed on this scaffold. Hence, the mason does not have to stoop over and pick up each brick from the floor of the platform on which he stands, thus saving time and energy.

The fountain trowel is a metal can shaped something like a low shoe. The heel is used to scoop up the mortar from the box, and the mortar is poured upon the brick through a narrow opening in the toe about 4 in. long. This fountain trowel makes it possible to spread a much greater quantity of mortar in a given time, and also permits the use of a softer mortar which fills the joints better.

170. Bond in Brick Masonry.—The bond is the arrangement of the brick in courses in such a way as to tie the wall together both longitudinally and transversely. The brick in one course should always break joints with those in the course below.

A stretcher is a brick laid with its greatest dimension parallel to the face of the wall, while a header is a brick laid with its greatest dimension perpendicular to the face of the wall. A course is a layer of brick, and is usually horizontal.

The three principal ways of bonding are the common, English, and Flemish methods. In the common bond, from four to seven courses of stretchers are laid to one course of headers. The English bond consists of alternate courses of headers and stretchers. In the Flemish bond, the headers and stretchers alternate in each course, and the brick are so placed that the outer end of a header lies in the middle of a stretcher in the course below.

Brick veneer consists of a single layer of brick placed on the face of the wall. Great care must be taken in the bonding of this veneer to the rest of the wall if these brick are to carry any of the load. One of the ways of bonding is a secret bond as shown in the sketch. Another way is to use metal ties extending from the joints in the veneer to the joints in the filling.

In hollow walls, the bonding of the outer layers to the inner layers is usually accomplished by long metal ties.

Arches in brick work are usually built with a series of header courses in which the brick are laid on edge. Such arches are called row-lock arches. Another method is to lay the brick with continuous radial joints with the brick laid partly as headers and partly as stretchers. Specially prepared brick must be used for this method.

171. Pointing of Brick Masonry.—After the wall is built, the edges of the exposed joints are pointed by refilling them to a depth of about one inch with specially prepared mortar. This mortar is usually richer than that used in building the wall. Sometimes the pointing mortar is colored so as to secure pleasing effects.

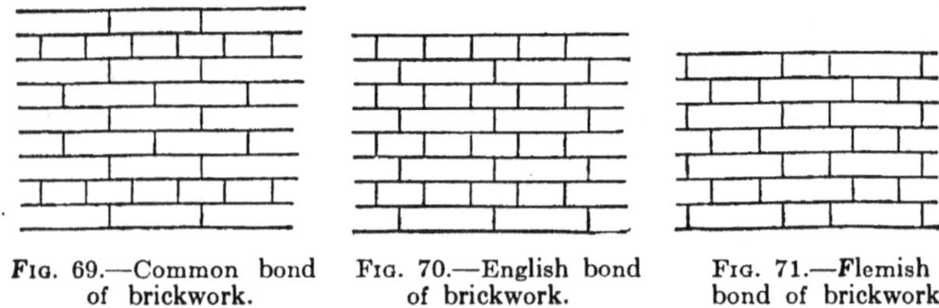

Fig. 69.—Common bond of brickwork.

Fig. 70.—English bond of brickwork.

Fig. 71.—Flemish bond of brickwork

The four general ways of refilling the horizontal joints are by making what are known as flush, bead, groove, and weather joints. The vertical joints are pointed in the same way except that when the horizontal joints are weather joints, the vertical ones are made flush. There are several other varieties of pointing which are not known by any general names. When the slope of the weather joint is reversed (sometimes called a struck joint), it allows water to collect in the joint and penetrate into the masonry.

172. Waterproofing Brick Masonry.—Brick masonry may be made water-tight by constructing it of impervious brick and mortar.

Another way is to incorporate a layer of tarred paper or felt (painted with asphalt or tar) in the wall.

If the wall is already built, it can be made more water-tight by painting it with a soap and alum solution, a tar or asphalt preparation, or some other waterproofing compound. Sometimes it is given a coating of impervious mortar or of an impervious bituminous mastic.

The methods of rendering a stone wall water-tight may be used to make a brick wall waterproof.

173. Cleaning Brick Masonry.—Brick masonry may be cleaned by the same methods as are used for cleaning stone masonry (see the article on "Cleaning Stone Masonry").

Mortar sometimes sticks so tightly to the brick that a metal tool is required to remove it.

Enameled brick can be cleaned with caustic soda or sodium carbonate, which does not have any effect on the brick or cement and lime mortar.

174. Strength and Other Properties of Brick Masonry.—The weight of the best pressed brick masonry with thin joints is about 145 lb. per cubic foot; of brick masonry of ordinary quality, 125 lb. per cubic foot; and of soft brick masonry with thick joints, 100 lb. per cubic foot. These values are approximate.

The strength of brick masonry depends more upon the strength of the mortar, the bond, and the workmanship than upon the strength of the brick. When it is desired to have strong masonry, a portland cement mortar must be used.

Occasionally, the transverse strength of the brick masonry is of importance, as in some cases the masonry acts as a beam (frequently when door openings are cut in a brick wall after it is built). A few tests have given results varying from 50 lb. per square inch to 300 lb. per square inch in cross bending, according to the quality of the brick and the mortar.

The modulus of elasticity of good brick masonry in compression is about 2,000,000 lb. per square inch.

The coefficient of expansion is approximately 0.0000030 per degree Fahrenheit for pressed brick masonry.

The shearing strength of brick masonry probably varies from 10 to 25 per cent of the shearing strength of the brick, depending upon the quality of the mortar used.

The compressive strength of brick masonry is of the most importance. A number of tests have been made on the crushing strength of brick masonry piers. The first sign of failure was usually a popping or cracking sound followed a little later by the appearance of cracks which gradually increased in size until the failure was complete. In nearly all of the tests, the mortar failed before the brick.

The following table gives results of compression tests made upon some brick piers at the Watertown (U. S.) Arsenal:

CRUSHING STRENGTH OF BRICK PIERS

Watertown Arsenal tests of 1904	Age 6 months					
Kind of brick	Compressive strength, pounds per square inch			Per cent of average crushing strength of the brick		
	Neat Portland cement	1 Portland 3 sand	1 lime 3 sand	Neat Portland cement	1 Portland cement 3 sand	1 lime 3 sand
Face brick						
Dry pressed face brick....	2,880*	2,400	1,517	26	21	13
Repressed mud brick.....	1,925	1,670	1,260	28	25	19
Common brick						
Wire cut stiff mud brick..	4,021	2,410*	1,420	31	19	11
Hard sand struck brick...	4,700*	1,800*	994	42	16	9
Hard sand struck brick...	1,969	1,800	733	44	40	16
Hard sand struck brick...	1,400	1,411	718	24	24	12
Light hard sand struck brick	1,510*	1,519	732	23	23	11
Light hard sand struck brick................	1,061	1,224	465*	20	23	9

*Age 1 month.

175. Allowable Working Loads for Brick Masonry.—Safe working loads for brick masonry in tension and cross bending should be considered as zero, except in special cases where there are special designs and constructions.

In the case of a lintel, the actual load on it is very uncertain. This load may be assumed to be the weight of all the masonry vertically above the lintel, including such loads as may be transmitted to the masonry from floors, etc. Another assumption is to take the load as the weight of the triangle of masonry above the lintel, considering the span of the lintel as the base of the triangle and assuming that the sides of the triangle make an angle of 45 degrees with the base. This latter assumption may be a little unsafe, but if the angle of the sides is changed to 60 degrees, the assumption gives results that are safe for most cases. If there is any doubt in regard to the safety of the lintel, it is better to construct an arch over the opening.

The allowable working stress in shear for brick masonry may be taken at one-fourth of the allowable working stress in compression.

The following table gives the safe working stresses in compression for brick masonry as recommended by a committee of

Chicago Engineers and Architects in 1908 for the building laws of that city. This table represents good practice.

SAFE WORKING LOADS IN COMPRESSION FOR BRICK MASONRY

Kind of brick	Kind of mortar		Safe load in pounds per square inch
Paving brick	1:3	Portland cement and sand	350
Pressed and sewer brick, strength 5,000 lb	1:3	Portland cement and sand	250
Select hard common brick, strength 2,500 lb	1:3	Portland cement and sand	200
Select hard common brick, strength 2,500 lb	1	Portland cement, 1 lime, 3 sand	175
Common brick, strength 1,800 lb. per square inch	1:3	Portland cement and sand	175
Common brick, strength 1,800 lb. per square inch	1:3	Natural cement and sand	150
Common brick, strength 1,800 lb. per square inch	1	Portland cement, 1 lime, 3 sand	125
Common brick, strength 1,800 lb. per square inch	1:3	Lime and sand	100

176. Efflorescence.—Efflorescence is a white deposit that frequently forms on the surface of brick masonry, especially in moist climates and in damp places, and spoils the appearance of the brickwork. The mortar in the masonry absorbs water and this water dissolves some of the salts of potash, soda, magnesia, etc. that are in the lime or cement. Then, when the water is evaporated from the surface of the brickwork, it leaves these salts in the form of a white deposit. Generally, there is a greater deposit from a lime mortar than from a natural or portland cement mortar. A portland cement mortar has the least amount of deposit. Sometimes the efflorescence originates in the brick, particularly if the brick were burned with sulphurous coal, or were made from clay containing iron pyrites.

Efflorescence can often be prevented by making the wall as water-tight as possible and by keeping water from leaking into the wall. Painting the wall with a soap and alum solution tends to prevent efflorescence.

Efflorescence can be removed from the wall by the use of scrubbing brushes and soap and water or a dilute solution of hydrochloric acid in water.

177. Hollow Tile Masonry.—Hollow tile masonry is composed of hollow terra cotta or tile building blocks laid in a portland cement mortar. This masonry is light in weight and fireproof. It is used for backing of walls, for entire walls, and for partitions and floors.

In walls, the blocks are usually laid with their openings vertical instead of horizontal. When the openings are vertical, wire

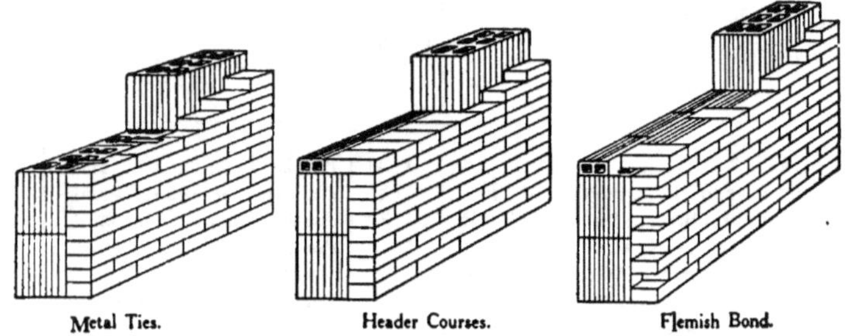

FIG. 72.—Hollow tile masonry wall veneered with brick.

screen is often laid in the horizontal joints to aid in holding the mortar in place. The mortar should be a 1:2 portland cement mortar, preferably containing a small amount of lime paste not to exceed ten per cent. Good hollow tile masonry can be safely used for the load-carrying walls of ordinary buildings and dwellings that are three stories or less in height.

Fireproof floors are often constructed of special hollow terra cotta blocks in buildings of steel or reinforced concrete construction. These blocks are set in portland cement mortar and are used as flat arches between the I-beam or reinforced concrete beam joists. A layer of concrete, about two inches thick, is usually placed on top of the blocks to form a wearing surface for the floor.

CHAPTER IX

TIMBER

A. TREES

178. Timber Trees in General.—Wood has long been used as a structural material because it could be obtained in most every locality and was easily adapted for use. While there are hundreds of varieties of trees, yet only a few species (probably less than 25 distinct species) are of great commercial importance.

Practically all of the woods used for structural materials are produced by the seed-bearing trees. These trees are divided into three groups: namely, the conifers, or soft woods (pine, spruce, fir, cedar, etc.); the broad-leaved trees, or hard woods (oak, maple, ash, walnut, hickory, etc.); and the tropical trees (bamboos, rattans, palms, etc.).

Of these three groups, the conifers, which are found throughout the northern hemisphere, are the most important structurally. The broad-leaved trees are found practically all over the world, and they are next to the conifers in structural importance. Of the soft- and hard-wood trees, probably the pine, fir, hemlock, spruce, cedar, oak, hickory, ash, poplar, maple, cypress, and walnut are the most important. Possibly the bamboo may also be classed as an important structural timber, especially in the tropical countries.

There is no sharp distinction in hardness between the soft woods and the hard woods, as some of the hard woods (such as poplar and basswood) are softer than some of the pines.

According to the manner of their growth, trees may be divided into two classes—the exogenous or outward growing trees (conifers and broad-leaved trees), and the endogenous or inner growing trees.

179. Structure of Exogenous Trees.—The structure of an exogenous tree consists of three parts—the bark, the sapwood, and the heartwood. The bark is a protective tissue found on the outside of the tree trunk and varying from one quarter to two inches in thickness. It is valueless as a structural material and is always removed soon after the tree is felled because it tends to hasten the decay of the wood. The sapwood is just inside of the bark and is made up of the soft thin-walled cells which form the living part of the tree. The heartwood, which is circular in shape and darker in color than the sapwood, is inside of the sapwood. The heartwood consists of many fibrous bundles which give the wood its strength and stiffness.

The wood of the exogenous trees is made up of bundles of long cells and fibers whose long axes are usually parallel to the tree

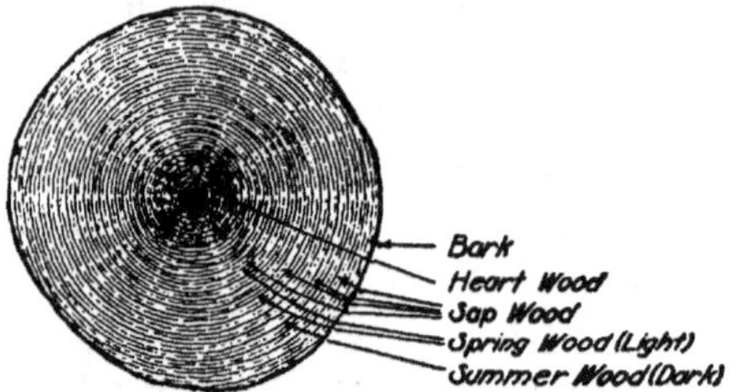

Fig. 73.—Cross-section of a tree showing annual rings.

trunk. These vertical bundles are crossed in a radial direction by plates of tissue or radial cells extending from the pith at the center of the tree to the soft tissue (sapwood) on the outside. These radial cells are called medullary rays and help to bind the longitudinal fibers more firmly together besides forming communications between the center of the tree and the outside. There are resin ducts scattered through the wood of the conifers and hollow ducts or vessels in the wood of the broad-leaved trees.

The conifers are more uniform in structure than are the broad-leaved trees, whose structure is often very complex.

180. Growth of Exogenous Trees.—The exogenous trees (conifers and broad-leaved trees) increase in size by the annual formation of new wood on the outer surface. The conifers can be recognized by their needle leaves, resinous bark, and cones, while the broad-leaved trees can be distinguished by their broad flaring leaves.

An exogenous tree grows in diameter when new and branching bundles of hollow fibers appear under the bark and form an annular ring on the outer edge of the sapwood. This happens once during each growing season, which extends through the spring and summer. The growth is more rapid in the spring than in the summer and this variation in growth causes a different appearance in the wood fibers, the summer wood usually being darker in color and denser than the spring wood. This makes the cross-section of the tree look like a number of concentric circular rings, each ring representing a year's growth. The age of a tree can be determined by counting the number of annular rings. The thickness of an annular ring varies from 0.01 to 0.5 in., with an average of about 0.10 to 0.15 in. The last few rings form the sapwood which is light in color and usually from ½ to 4 in. thick. The rings inside the sapwood form the heartwood which contains from 25 to 85 per cent of the wood of the tree, according to the kind of tree and the conditions of growth. The time required for the sapwood to change to heartwood varies from a few years in the fir to many years in the oak.

The exogenous trees grow in length because each annular layer extends over the others, thus increasing the length. Because of the conical shape of the tip, the increase in length may be much greater than the increase in diameter.

Knots in a tree are caused by the encasement of a limb by the successive annual layers of wood. When a board is sawed out of a tree, the knot is the portion of the branch contained in the board, and the fibers of the knot are usually about perpendicular to the other fibers in the board. A loose knot is one that is loose or badly cracked or checked so as not to be solid in the board, and it is usually composed of dead wood. A sound knot is one that is solid and contains no appreciable cracks or checks. It is usually composed of living wood.

181. Structure and Growth of Endogenous Trees.—These trees are largely confined to the tropical regions, and the palms and bamboo are about the only ones of structural importance.

The wood elements of endogenous trees are similar to those of exogenous trees but their arrangement is different. The fibrous bundles do not form concentric circles around the center of the tree, but are scattered throughout the wood where they curve inward and outward among each other thus making a more complex structure.

Endogenous trees increase in diameter and length by the intermingling of new wood fibers with the old. The growth of the fibers is apt to be more rapid in the outer part of the stem, thus causing the outer part to be more dense and solid than the inner. When the growth is very rapid, a hollow is formed in the center of the stem, because of the insufficient growth and the rupture of the inner fibers. Knots or joints are often found at the places where leaves have issued. The bamboos have hollow centers, while the palms and yuccas have pithy centers.

182. Grain and Texture of Wood.—Depending on the character and arrangement of wood elements, the width of growth

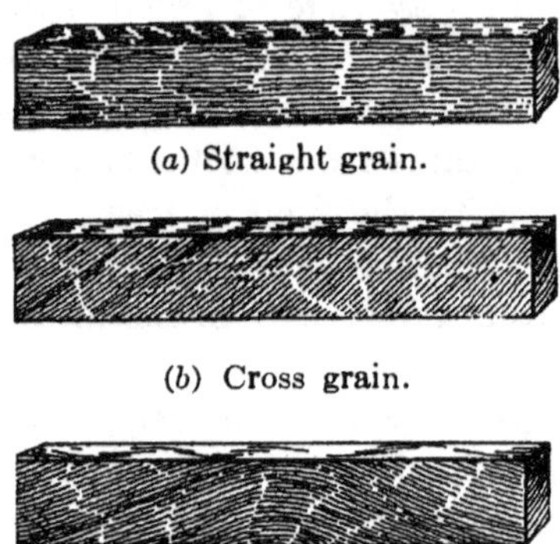

(*a*) Straight grain.

(*b*) Cross grain.

(*c*) Twisted grain.

FIG. 74.—Showing the grain of wood.

rings, etc., wood may be described as fine or coarse grained, straight or twisted or cross grained, curly, "bird's-eye," or mottled grained, etc.

Woods are fine grained if their growth rings are narrow, and coarse grained if their growth rings are wide. Woods may be said to be rough or smooth grained according to the appearance of the surface. They are straight grained if the fibers are straight and parallel to the axis of the tree; twisted if the fibers follow a spiral course around the tree; cross grained if the fibers change direction during the growth; curly grained if the fibers tend to form short curves or curls (as in curly birch); mottled if the appearance has a mottled effect. "Bird's-eye" is probably due to the layer of wood next to the bark becoming pitted or marked by small projections, probably caused by the presence of undeveloped buds as in "bird's-eye" maple.

Woods may be said to have coarse or fine texture if the elements are large or small. The texture is even if the fibers are all of about the same size, and uneven if the size varies.

183. Color and Odor of Wood.—Color helps in identifying the species of wood. Most new wood is almost colorless but becomes yellowed after a few years and usually deepens in color when the sapwood changes to heartwood. The color may be variable or uniform throughout the heartwood and may be lighter or darker according to the species and the manner of growth. Deep color is nearly always due to the infiltration of resins, pigments, tannins, etc. into the heartwood. All woods darken more or less when exposed for a time to air or immersed in water. Hence, the natural color can only be observed in newly cut wood. Color aids in distinguishing the heartwood from the sapwood, as the heartwood is nearly always darker than the sapwood.

All woods possess a characteristic odor, though in some cases it is not readily distinguished. The odor is due to foreign chemical compounds in the wood and is usually more pronounced in heartwood than in sapwood. The odors of green wood, seasoned timbers, and decaying wood are different in different species and aid in identifying the different species. A few of the woods lose most of their odor when they are seasoned.

184. General Characteristics of Conifers—Pine, Fir, Spruce.—*White Pine.*—Light, soft, straight grained, easily worked, but not very strong. Light yellowish brown color often tinged slightly with red. Used for pattern making and interior finishing.

Red Pine (Norway Pine).—Light, hard, coarse grained, compact, with few resin pockets. Light-red color with a yellow or white sapwood. Used for all purposes of construction.

Yellow Pine (Long Leaf).—Heavy, hard, strong, tough, coarse grained, and very durable when dry and well ventilated. Cells are dark colored and very resinous. Color, light yellowish-red or orange. Cannot be used in contact with the ground, as it then decays rapidly. Used for heavy framing timbers and floors.

Yellow Pine (Short Leaf).—Varies greatly in the amount of sap and quality. Cells are broad and resinous with numerous large resin ducts. Medullary rays well marked. Color, orange with white sapwood. Used as a substitute for long leaf pine.

Douglas Fir (Oregon Fir).—Hard and strong but varying greatly with age, conditions of growth, and amount of sap. Durable but difficult to work. There are two varieties, red and yellow, of which the red is the more valuable. Color, light red to yellow with a white sapwood. Used in all kinds of construction.

Black Spruce.—Light, soft, close grained, straight grained, and satiny. Color, light red and often nearly white. Resists decay and the destructive action of crustacea. Used for piles, framing timbers, submerged cribs, and cofferdams.

White Spruce.—Similar to black spruce, but it is not so common. Light-yellow in color with an indistinct sapwood. Used for lumber in construction work.

185. General Characteristics of Conifers—Other Species.—*Hemlock.*—Soft, light brittle, easily splits. Is not durable, is likely to be shaky, and has a coarse uneven grain. Light brown color tinged with red, and often nearly white. Resistant to attacks of ants. Used for cheap, rough, framing timber and some finishing lumber.

White Cedar.—Soft, light, fine grained, and very durable in contact with the soil. Lacks strength and toughness. Light-brown color which darkens with exposure. Sapwood is very thin and nearly white. Used for water tanks, shingles, posts, fencing, cooperage, and boat building.

Red Cedar.—Strong pungent odor, repellant to insects. Very durable and compact, brittle, but easily worked. Color, dull-brown tinged with red. Used for posts, sills, ties, fencing, shingles, and linings for chests, trunks, and closets.

Tamarack (Larch).—Hard, heavy, strong, durable. Like spruce in structure and hard pine in weight and appearance. Used for posts, poles, sills, ties, and ship timbers.

Cypress.—Very durable, light, hard, close grained, brittle, easily worked, and polishes easily. Color, bright clear yellow with a nearly white sapwood. Used for house siding, building lumber, poles, interior finishing, etc. Resists dampness and excessive heat.

Redwood (California or Giant).—Light, soft, weak, and brittle. Grain is coarse, even, and straight. Easily split and worked. Durable when in contact with the soil. Shrinks lengthwise as well as crosswise. Color, bright clear red becoming darker with exposure. Used for ties, posts, poles, and as a general building material.

186. General Characteristic of Broad-leaved Trees—Oak, Maple, Ash, Walnut.—*White Oak.*—Heavy, strong, hard, tough, and close grained. Checks if not carefully seasoned. Well-known silver grain. Capable of taking a high polish. Color, brown with lighter sapwood. Used for framed structures, shipbuilding, interior finish, carriage, and furniture making.

Chestnut Oak.—A species of white oak. Very durable in contact with the soil. Dark-brown color. Used for ties.

Live Oak.—Very heavy, hard, tough, and strong. Hard to work. Color, light-brown or yellow with a nearly white sapwood. Used in shipbuilding and wagon work.

Red and Black Oak.—More porous than white oak and softer and less strong. Color, darker and redder than white oak. Used for furniture and interior finish.

Hard Maple.—Heavy, hard, strong, tough, and coarse grained. Medullary rays are small but distinct. Easy to polish. Color, very light-brown to yellow. Used for flooring, interior finish, and furniture.

White Maple.—About the same as hard maple except that it is lighter in weight and color. Same uses.

White Ash.—Heavy, hard, very elastic, coarse grained, and compact. Tends to become decayed and brittle after a few years. Reddish brown color with a nearly white sapwood. Used for interior finish and cabinet work. Unfit for structural work.

Red Ash.—Heavy, compact, and coarse grained but brittle. Color, rich-brown, with sapwood a light-brown sometimes streaked with yellow. Used as a substitute for the more valuable white ash.

Green Ash.—Heavy, brittle, hard, and coarse grained. Color, brown with lighter sapwood. Used as a substitute for white ash.

White Walnut (Butternut).—Light, soft, coarse grained, compact, and easily worked. Polishes well. Color, light-brown turning dark on exposure. Used for interior finish and cabinet work.

Black Walnut.—Hard, heavy, strong, and coarse grained. Checks if not carefully seasoned. Easily worked. Rich dark-brown color with a light sapwood. Used for interior finish and furniture.

187. General Characteristics of Broad-leaved Trees—Other Species.—*White Elm.*—Heavy, hard, strong, tough, and very close grained. Difficult to split and shape. Warps badly in drying. Takes a high polish. Color, light-clear-brown often tinged with red and gray, with a broad whitish sapwood. Used for building cars, wagons, boats, and ships. Used for sills, bridge timbers, ties, furniture, and barrel staves.

Hickory.—Heaviest, hardest, toughest, and strongest of the American woods. Very flexible. Medullary rays numerous and distinct. Brown in color with a valuable white thin sapwood.

Used for carriages, handles, and bent wood instruments. Not used for structural purposes on account of its hardness and liability to attack by boring insects.

Locust.—Heavy, hard, strong, and close grained. Very durable in contact with the ground. Hardness increases with age. Color, brown (and rarely light-green) with yellow sapwood. Used for ties, vehicles, posts, and turned ornaments.

Gum.—Heavy, hard, tough, and close grained. Shrinks and warps badly in seasoning. Not durable when exposed to weather. Takes a high polish. Color, bright-brown tinged with red. Used for furniture, hat blocks, wagon hubs, interior finish.

Mahogany.—Strong, durable, and flexible when green, and brittle when dry. Free from shakes. Not very liable to attacks of dry rot and worms. Peculiarly marked by short straight lines or dashes. Rapid seasoning causes deep shakes. Color, red-brown of various shades and often varied and mottled. Used for interior finish, furniture, veneers, etc.

Chestnut.—Light, moderately soft, stiff, and of coarse texture. Shrinks and checks considerably when drying. Easily worked. Durable when exposed to the weather. The heartwood is dark and the sapwood light-brown in color. Used for cabinet work, cooperage, ties, telegraph poles, and exposed heavy construction.

Poplar (Whitewood).—Soft, very close and straight grained, brittle. Shrinks excessively in drying. Warps and twists very much but does not split when dry. Easily worked. Light-yellow to white color. Used for vehicles, wooden instruments, toys, furniture, finishing, etc.

Lignum-Vitæ.—Very hard, heavy, resinous, has a soapy feeling, and is difficult to split and work. Color, rich yellow-brown varying to almost black. Used for small turned articles, tool handles, and sheaves of block pulleys.

Teak.—Tropical wood, durable, heavy, hard, elastic, strong, and easy to work. When seasoned it does not rack, split, shrink, or alter in shape. Aromatic odor. Heartwood is golden-brown in color, seasoning into brown. The sapwood is white. Used in temples, ships, buildings, and for structural timbers. Can be used in contact with iron.

Catalpa.—Light weight, soft, weak, elastic, and durable in contact with the soil. Used for ties, posts, cabinet work, and interior finishing.

Eucalyptus.—Very hard, heavy, strong, tough, and close

grained. Hard to split after it is dried. Not durable in contact with the soil. Resembles ash and hickory in appearance. Resists attacks of marine borers. Used for wharf piling, lumber, parts of vehicles, furniture, flooring, paving, etc.

Beech.—Hard, heavy, strong, and tough. Not durable when exposed. Subject to attack by insects. Liable to check in seasoning. Takes a high polish. Color, white to light-brown or reddish. Used for furniture, interior finish, ship building, and carriage making.

188. General Characteristics of Some Endogenous Trees. *Palmetto.*—Lightweight. Difficult to work when dry. Very durable under water as it resists attacks by the Teredo and borers. Color, light-brown with dark-colored fibers. Used for piles and wharves.

Bamboo.—Hollow stem with many joints. Many branches, usually small ones. Used for timbers, columns, masts, poles, rafters, water pipes, furniture, split bamboo work, etc.

B. PREPARING THE TIMBER

189. Logging.—Logging may be said to be the process of felling the trees, trimming off the branches and vegetation, cutting the trunks and limbs to proper sizes, and transporting the logs to the sawmill. The trees are felled by means of axes or saws and are then chopped or sawn into sizes small enough to be transported to the sawmill. The methods of transportation vary according to conditions. If the sawmill is quite close to the forest, the logs may be rolled downhill, placed on sleds drawn by horses, pulled by a donkey engine and windlass, drawn by automobiles, carried by an aërial tramway, floated down small streams, or carried by a narrow-gage logging railway to the mill. If the sawmill is some distance away, the logs may be transported by railway or assembled in rafts and floated on a river or other waterway to the mill.

It is important to choose the proper time for cutting the timber. In the spring and late summer, the sapwood contains an abundance of moisture with starches, sugars, and oils in solution, all of which tend to hasten the decay of the timber. In the drier summer months and in the winter, the growing and conducting cells of the tree are less active or altogether dormant, and the best wood is secured if the timber is cut during those seasons. Oak is claimed to be more durable if it is cut just after the leaves have fallen. Hewn lumber is thought to be more durable than sawn lumber. Usually, most of the logging operations are carried on during the winter months.

190. Sawing the Lumber.—Most of the sawing of lumber is done in sawmills by machine driven rotary or band saws. The manner in which a stick of lumber is sawed from the log has a

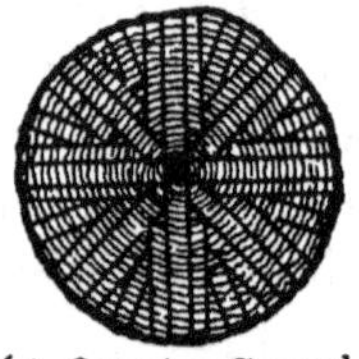

(*a*) Flat Sawed (*b*) Quarter Sawed (*c*) Quarter Sawed

Fig. 75.—Methods of sawing lumber.

remarkable influence on its qualities and behavior. The kind of sawing is determined by the character of the wood and the purpose for which it is to be used.

The two main classes of sawing are flat and rift sawing. Flat sawing consists in cutting the timber tangential to the annular rings. Rift (or quarter) sawing is cutting the boards out of the log in such a manner that the annular rings are cut through as nearly as possible in a radial direction. Quarter sawing is done for the sake of the beauty of the grain thus obtained, as well as to expose the edge of the hard bands of the summer wood. Flat sawing and rift sawing give rise, in the lumber trade, to the terms flat grain and edge grain respectively. Edge grain lumber does not sliver, shrinks and checks less, and wears more evenly and smoothly than the flat grain lumber.

Ordinary, or bastard, sawing consists of cutting the log into a number of parallel slices and then trimming the edges of these slices with a circular saw. In ordinary sawing, some of the boards will be flat sawn, some quarter sawn, and about half of them will be neither flat nor quarter sawn but a combination of these.

191. Classification of Lumber.—All material sawn from logs for structural or other commercial purposes is called lumber. The larger sizes, such as beams, joists, etc., are called timbers, and these timbers are usually resawn in order to obtain the smaller sizes of lumber. Lumber is furnished in all sizes and dimensions such as are suitable for the work at hand.

The term "resawed lumber" is applied to lumber sawed on all four sides. Rough edge or flitch is lumber sawn on two sides. Planed resawed lumber is called dressed lumber. Dressed planks and boards free from all defects are called clear. Such boards are produced in regular sizes ⅛ in. less in thickness than the sawed lumber, and ranging from ⅝ to 1⅞ in. in thickness.

Sawed timbers shall be sound, of standard size, square edged, and straight; and they shall be close grained and free from defects, such as injurious ring shakes and cross grain, unsound or loose knots, knots in groups, decay, or other defects that will materially impair the strength.

Rough sawing to standard size shall mean that the timbers shall not be over ¼ in. scant from the actual size specified; for instance, a 12 by 12 timber shall not measure less than 11¾ by 11¾ in.

Standard dressing shall mean that not more than ¼ in. shall be allowed for dressing each surface; for instance, a 12 by 12 timber after being dressed on four sides shall not measure less than 11½ by 11½ in.

The standard lengths are multiples of 2 ft., running from 10 to 24 ft. for boards, fencing, dimension, joists, and timbers. Longer or shorter lengths than those herein specified are special lengths. Special and fractional lengths shall be counted as of the next higher standard length.

The standard widths for lumber shall be multiples of 1 in.

All sizes 1 in. or less in thickness shall be counted as 1 in. thick.

Flooring shall include pieces 1, 1¼, and 1½ in. thick by 3 to 6 in. wide, excluding 1½ by 6.

Boards shall include all lumber less than 1½ in. thick and more than 6 in. wide.

Plank shall include all sizes from 1½ to under 6 in. in thickness by 6 in. or over in width.

Scantling shall include all sizes exceeding 1½ in. and under 6 in. in thickness, and from 2 to under 6 in. in width.

Dimension sizes shall include all sizes 6 in. and more in thickness by 6 in. and more in width.

Stepping shall include all sizes from 1 to 2½ in. in thickness by 7 in. and over in width.

Rough edge, or flitch, shall include all sizes 1 in. and more in thickness by 8 in. and more in width, sawed on two sides only.

192. Defects in Lumber.—The following defects are adopted as standard by the American Society for Testing Materials. The diameters of the knots and holes are average diameters.

Knots.—A sound knot is one which is solid across its face and which is as hard as the wood surrounding it; it may be either red or black, and is so fixed by growth or position that it will retain its place in the piece of lumber.

A loose knot is one not held firmly in place by growth or position.

A pith knot is a sound knot with a pith hole not more than 1/4 in. in diameter at the center.

An encased knot is one which is surrounded wholly or in part by bark or pitch. Where the encasement is less than 1/8 of an inch in width on both sides, not exceeding 1/2 the circumference of the knot, it shall be considered a sound knot.

A rotten knot is one that is not so hard as the wood it is in.

A pin knot is a sound knot not over 1/2 in. in diameter.

A standard knot is a sound knot not over 1 1/2 in. in diameter.

A large knot is a sound knot more than 1 1/2 in. in diameter.

A round knot is one which is oval or circular in form.

A spike knot is one sawed in a lengthwise direction.

Wane.—Wane is bark, or the lack of wood from any cause, on edges of timbers.

Pitch Pockets are openings between the grain of the wood containing more or less pitch or bark. These shall be classified as small, standard, and large pitch pockets. A standard pitch pocket is one not over 3/8 of an inch wide or 3 in. in length. A small pitch pocket is one not over 1/8 of an inch wide. A large pitch pocket is one over 3/8 of an inch wide, or more than 3 in. long.

A Pitch Streak is a well defined accumulation of pitch at one point in the piece. When the pitch is not sufficient to develop a well defined streak, or where the fiber between grains (the coarse grained fiber, usually termed "spring wood") is not saturated with pitch, it shall not be considered a defect.

Shakes are splits or checks in timbers which usually cause a separation of the wood between the annual rings. A ring shake is an opening between the annual rings. A through shake is one which extends between two faces of a timber.

Rot, Dote, and *Red Heart* are forms of decay which may be evident either as a dark red discoloration not found in sound wood, or by the presence of white or red rotten spots, and shall be considered as defects.

193. Natural Seasoning of Lumber.—In the preparation of lumber for construction purposes, it is necessary to expel the sap and moisture from the pores of the wood by some natural or artificial means. This process is called seasoning. It has been found that the drier the timber, the less likely it is to shrink and decay.

Natural air seasoning consists in exposing the planks and boards, after sawing, to a free circulation of air. The lumber is placed on skids in large square piles under shelter in a dry place, the layers being separated by three or four narrow strips or boards laid in the opposite direction. The lowest layer should be at least 2 ft. from the ground. At frequent intervals the decayed pieces should be removed and the lumber repiled. The time required for thorough seasoning varies from 1 to 3 years, depending upon the character of the wood, the purpose for which it is to be used, and its dimensions.

Water seasoning is another type of natural seasoning which consists in immersing the lumber in water. The soluble substances in the sapwood are removed, leaving a timber that is less liable to warp and crack. Water seasoning causes the heartwood to become brittle and lose its elasticity. In this method of seasoning, the timber is immersed for about 2 weeks and then removed and thoroughly dried with an excess of air. If immersed too long, the wood becomes brashy when exposed to the air. Water seasoning is not very much used.

194. Artificial Seasoning of Lumber.—Artificial seasoning or kiln drying hastens the evaporation of the moisture and the removal of the sap, but it produces an inferior product because it causes a rapid drying of the surfaces and ends of the material and a slow or imperfect drying of the interior. This weakens both the strength and the elasticity of the wood.

The timber is stacked in a drying kiln and exposed to a current of hot air, the temperature depending upon the kind of lumber and its dimensions. Sometimes vacuum pumps are used in connection with the heating. The temperature usually varies from about 100 degrees Fahrenheit for oak to about 200 degrees Fahrenheit for pine. The time required depends upon the thickness of the lumber. About 4 days are required for 1 in. pine, spruce, or cedar boards. Hard woods are usually dried in air from 3 to 6 months and then placed in the drying kiln from 6 to 10 days.

When rapidly dried in a kiln, oak and other hard woods tend to become "case-hardened" as the outer parts dry and shrink before the interior parts have a chance to do the same. Thus there is a firm shell of dry, shrunken, and usually checked wood about the interior. When the interior drys, it tends to become checked along the medullary rays. Lumber that has been properly air dried will not case-harden when placed in a kiln.

195. Shrinkage of Lumber.—When a short piece of wood fiber dries, it shrinks; its walls become much thinner and the cavity becomes greater, but the length of the fiber remains about the same. A thick-walled fiber shrinks more than a thin-walled one. As most of the fibers in a tree are parallel to its length, the length of a timber will not change appreciably, but the cross section will shrink when the timber is seasoned. The medullary rays have an effect on the shrinkage of the cross-section, the wood in the cross-section shrinking more at right angles to the rays than parallel to them, due to the fact that the rays themselves shrink in cross-section but not in length. Hence, the greatest shrinkage in lumber will take place tangentially to the annular rings, a little less shrinkage will take place in a direction radially to the annular rings, while the shrinkage in the longitudinal direction of the tree will be inappreciable. The shrinkage of the fibers tangentially to the annular rings is known as circumferential shrinkage. Some woods shrink much more unevenly than others. The harder timbers are more compact in structure, with thicker cell walls, and, therefore, produce the greatest shrinkage.

Quarter-sawed lumber will shrink less than flat-sawed lumber. A combination of quarter and flat sawing will cause the lumber to shrink unevenly, thus causing a warped surface. Flat sawing produces lumber that checks and cracks to a greater extent in drying than rift sawed lumber does.

If the outer fibers of a timber dry out much faster than the inner fibers do, the timber will tend to become checked and cracked. This tendency toward checking and cracking may be reduced by driving S-irons, etc. in the ends of the timbers.

If a board shrinks unevenly, it will become warped. This may be due to the fibers on one side drying out faster than the ones on the other side, uneven drying, sawing in such a way that the shrinkage will be more in some directions than in others, or due to the structure of the board itself.

The opposite effect to shrinkage is produced by the absorption of moisture, and precautions must be taken when applying timber to construction work to allow for this expansion, such as expansion joints in a wood-block pavement. A roadway 40 ft. wide (constructed of wood blocks) has been observed to expand 8 in.

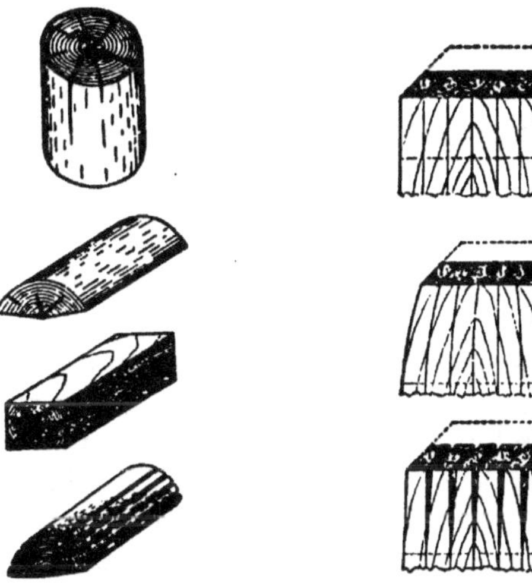

Fig. 76.—Effects of shrinkage. (*Bull.* 10, *U. S. For. Div.*)

Fig. 77.—Formation of checks. (*Bull.* 10, *U. S. For. Div.*)

The longitudinal shrinkage of timber is usually less than $\frac{1}{10}$ of 1 per cent. The change in volume of the timber is due to the radial and tangential shrinkage, and expressed in percentages is approximately twice the figures given in the following table, as the shrinkage takes place in two directions in approximately equal amounts. The following are average values for shrinkage in width:

Shrinkage of the Width of Wood

Kind of Wood	Per Cent Shrinkage
Light conifers (soft pines, spruce, cedar, cypress)...	3
Heavy conifers (hard pine, tamarack).............	4
Honey locust, box elder, wood of old oaks.........	4
Ash, elm, walnut, poplar, maple, beech, cherry, sycamore....................................	5
Basswood, birch, chestnut, blue beech, young locust	6
Hickory, young oak (especially red oak)...........	up to 10

C. DURABILITY AND DECAY OF LUMBER

196. Durability and Decay of Lumber in General.—The life of timber depends upon the way in which it is felled, seasoned, and worked. The timber is subject, in both its growing and converted states, to decomposition and attack by animal and vegetable life. Trees should be felled when the growing and conducting cells are less active or are dormant. Seasoning increases the life of timber by removing the sap and moisture. In structural work the timber should be protected as much as possible from the attack of agencies that cause decay.

The agencies which produce the decay of wood are: alternate moisture and dryness, heat and confined air, bacteria and fungi, insects and worms. Well seasoned wood in a uniform state of moisture or dryness and well ventilated should never decay. Timber that is kept constantly immersed in water may soften or weaken but it will not decay. Elm, elder, oak, and birch possess great durability when kept constantly immersed.

Dryness and ventilation are the best preventives of the decay of timber used for construction purposes. Wood that has been kept dry has been known to last for hundreds of years, though it finally became brittle and lost most of its strength. In construction work it is important that timber be kept completely immersed in water or else kept in a fairly dry condition and well ventilated. Water should be prevented from collecting in the joints, and important structural timbers should be protected from weather conditions when practical.

197. Dry Rot in Lumber.—Dry rot is directly caused by the fermentation and breaking down of the chemical compounds of the wood, due to the introduction of a certain fungus in the presence of a little moisture. These lower organisms excrete ferments which dissolve out parts of the cell walls, thus causing a crumbling of the wood. The growth of this fungus is stimulated by moderate warmth, presence of dampness, and lack of ventilation. Dry rot is often found in ill-ventilated places, such as the wall pockets at the ends of floor timbers, and in the core of timber columns in mill construction. The decomposition is often hastened by the use of unseasoned wood.

Dry rot is indicated by a swelling of the timber and a change in the color, the wood gradually becoming covered with mold and emitting a musty odor. Sometimes reddish or yellowish spots appear on the timber, and the fibers are gradually reduced to a powder. Dry rot is especially dangerous as it destroys the timber in which it originates and also tends to spread to adjacent woodwork. It is difficult to eradicate, when it is once established, the only remedy being to remove all of the fungus and disinfect the wood. Actual contact is not necessary for the spreading of dry rot.

198. Wet and Common Rot.—Wet rot appears only when the wood is kept damp or is subject to alternate dryness and moisture. It will not take place if the wood is thoroughly seasoned and the further absorption of moisture prevented. The decay is caused by the moisture which dissolves out the substance of the cell walls of the sapwood. Wet rot spreads by actual contact only. Wood cut in the spring and early fall is especially subject to wet rot. The remedy is to remove all of the rotten parts of the timber and keep the remainder dry and well ventilated.

Common rot is shown by the presence of external yellow spots on the ends of timber sticks and often by a yellowish dust in the checks and cracks, especially where the timbers are in contact with each other. The cause of common rot is improper seasoning in badly ventilated sheds.

199. Injurious Insects.—The larvæ of many insects are destructive to wood. The living trees are attacked by some, and the felled trees and lumber by others.

Some of the common insects attacking the wood of living trees are the oak and chestnut timber worms, locust borers, carpenter worms, ambrosia beetles, turpentine beetles and borers, and the white pine weavil.

Round timbers with the bark on are subject to attack by the insects mentioned above, and especially by the round-headed borers, timber worms, and ambrosia beetles.

Seasoned and finished hardwood lumber is especially subject to attack by powder post beetles.

Construction timbers are often seriously injured by wood boring larvæ, termites, black ants, carpenter bees, and powder post beetles.

The damage is caused by the insects, or their larvæ, eating or "boring" holes in the timbers, thus breaking the continuity of the fibers and reducing the cross-sectional areas.

200. Marine Wood Borers.—The Teredo or ship worm belongs to the mollusk species and is the marine borer which is the most active, and destructive to wood. It bores its way in lumber usually in a direction parallel to the grain and lines the hole with a calcareous deposit as it progresses. These worms vary much in size, some of the largest being about half an inch in diameter and 4 or 5 ft. long. They live in clear salt water, preferably of the warmer climates, and are more active near calcareous shores. They work from the ground up to the half-tide level.

The lycoris fucata is a little worm with many legs something like a centipede. It crawls up the piles or timbers inhabited by the Teredo, enters the hole, finds and eats the Teredo, and then lives in the hole.

The xylotrya also belongs to the mollusk species, and is similar to the Teredo in structure and mode of life.

The limnora, or gribble, is a small crustacean resembling a wood louse and is about the size of a grain of rice. It can swim, crawl, and jump. Both air and water are required for its existence; consequently, its attacks on wood are confined to a space between the high- and low-water marks. It devours the wood at the rate of 1 to 3 in. a year, and is found in both warm and cold water.

D. PROPERTIES OF TIMBER

201. Strength of Timber in General.—The mechanical properties of wood are very variable, not only between different kinds of trees, but between trees of the same kind, and even between specimens cut from different parts of the same tree. In estimating the properties of timber the following things should be considered—correct identification of the species and variety, age and rate of growth of the trees, position of test specimen in the tree, moisture content, and freedom of test specimens and commercial timbers from defects.

In general, the results of tests have shown that:

The influence of defects is very marked. Defects tend to lower the ultimate strength. Knots and cross grains lower the elastic limit.

Tests on small specimens usually give results that are at least 50 per cent greater than the results obtained from tests on large specimens.

Large checks and seasoning cracks weaken the wood.

In general, the strength of wood varies with the specific gravity.

Timber treated with creosote, tannin, zinc chloride, etc. is usually weaker than untreated timber.

Dry timber is much stronger (about 75 per cent) than wet or green timber.

The strength parallel to the grain is different than the strength perpendicular (across) to the grain.

The strength of timber under any kind of a permanent load is only about one-half of the strength found by short time tests.

Rapid loading in tests will give higher results than slow loading.

In general, the larger the percentage of summer wood, the stronger the timber.

In general, the strength of wood varies with the number of annular rings per inch.

It must be remembered that the percentage of moisture is the greatest factor influencing the strength of timber; hence, the percentage of moisture in the specimens tested should always be given.

202. Influence of Moisture Content in Timber.—Moisture has a very great influence upon the strength of timber, probably more than any other factor. The strength and weight of timber depend, to a large extent, upon the number of fibers per unit area of cross section; hence, the more fibers per unit area the heavier and stronger the wood. Absorption of moisture by the wood causes the fibers to swell in diameter and thus the number of fibers per unit cross sectional area are decreased and the wood is not so strong. Further, moisture tends to weaken the cells and make them less firm and strong. Results of tests have shown that, in the seasoning of Southern pines from green wood (33 per cent of moisture) to dry wood (about 10 per cent of moisture), there were variations of over 75 per cent in the average strength, the strength increasing with the decrease in moisture.

The strength decreases with increase in the moisture content up to the point where the cell walls become completely saturated. This limit is between 20 and 30 per cent for most woods. The addition of more moisture to the wood fills the cavities and causes no further swelling of the cell walls and has practically no effect upon the strength.

The amount of moisture contained in ordinary dry lumber is about 15 per cent, and this value varies greatly with the temperature and weather. So-called "dry" wood usually has as much as 8 per cent of moisture, while green and wet woods contain over 30 per cent. It is practically impossible to obtain perfectly dry wood. Wood is said to be dry when it has been dried to a constant weight (the variation in weight for a period of

24 hours being less than $\frac{1}{2}$ of 1 per cent) in an oven where the temperature was kept approximately at 212 degrees Fahrenheit.

203. Tensile Strength of Timber.—The tensile strength of timber is not of much importance except as it is involved in transverse loading. In construction, timber is rarely ever subjected to pure tensile stresses, due to the difficulty of designing proper end fastenings.

Failure in tension across the grain is due to the tearing or pulling apart of the wood fibers longitudinally. The tensile strength across the grain is only a small part ($\frac{1}{10}$ to $\frac{1}{20}$) of the tensile strength parallel to the grain.

Failure in tension along the grain is due to the transverse or oblique tearing apart of the wood fibers. That is, the fibers are rarely pulled in two, but they are usually pulled out from between the others. Knots, cross grain, medullary rays, and other defects weaken the timber in tension.

The proportional elastic limit of wood in tension along the grain is usually between 60 and 75 per cent of the ultimate strength.

204. Compressive Strength of Timber.—The compressive strength of timber is important as timbers are very frequently used as columns and compression members in various structures.

In compression along the grain, the fibers act like a number of hollow columns bound together. When failure occurs, the fibers tend to bend or buckle over each other and shear off. The compressive strength along the grain depends upon the density of the wood, the stiffness and continuity of the wood fibers, adhesion between the fibers, seasoning, moisture content, straightness of grain, and defects.

The proportional elastic limit in compression along the grain is usually between 60 and 75 per cent of the ultimate strength.

In compression across the grain the fibers fail by flattening. The strength depends primarily on the density of the wood, though many of the other factors have some influence. The strength across the grain is from $\frac{1}{6}$ to $\frac{1}{4}$ of that along the grain.

205. Transverse Strength of Timber.—The transverse strength of timber is important as timbers are often used as beams and joists in structural work.

The strength of timber in cross bending depends largely upon the compressive, tensile, and shearing strengths. Consequently, the transverse strength of timber depends upon the same factors as the tensile and compressive strengths do.

Timber beams rarely have a final failure in compression, though the initial failures are nearly always in compression. Final tension failures are quite common, especially if the beam is long compared with its thickness or if there are defects on the tension side. A large proportion of the failures occur by horizontal shear, especially if the beam is short and thick.

The "elastic limit" in cross bending is usually between 66 per cent and 75 per cent of the ultimate strength. A load in excess of the elastic limit will cause a beam to break if left loaded.

The modulus of elasticity in cross bending is a variable quantity because the load deflection curve is not a straight line, even as far as the elastic limit. The value of the modulus of elasticity in cross bending is about the same as that in compression.

Stiffness is the ability of a beam to resist cross-bending loads without having large deflections. The transverse modulus of elasticity may be considered as a measure of the stiffness. Straight grained lumber is stiffer than knotty or cross-grained pieces, and dry wood is about one and a half times as stiff as green or wet wood. In general, the heavier the wood, the stronger and stiffer it is.

206. Shearing Strength of Timber.—The resistance to shear across the grain is from four to ten times that along the grain. The shearing strength is less in wet woods than in dry woods, and it is reduced by defects, such as knots, checks, cracks, etc. The shearing strength along the grain is very small and depends upon the adhesion of the wood fibers to each other, the straightness of grain, the medullary rays, etc. The shear across the grain is approximately half of the compressive strength along the grain and depends upon the same factor.

The shearing strength of timber is important, especially in beams.

207. Cleavability and Flexibility of Timber.—Cleavability is the resistance of wood to splitting, as with an axe or other tool. Elastic woods split more easily than others, while woods with great hardness and transverse tensile strength are hard to split.

Wood splits naturally along the two normal planes and a little more readily along the radius. The weight of the wood has but little effect on the cleavage. Defects, especially knots and cross grains, and the presence of moisture, increase the resistance of the wood to splitting.

Flexibility is the ability of the wood to bend very much without breaking. Hard wood is usually more flexible than soft wood. Moisture softens the wood and makes it more flexible, while knots and other defects make it less flexible.

208. Hardness and Toughness of Timber.—Hardness is usually measured by the penetration of a ball or a steel plunger under a load. The hardness is closely related to the shearing strength across the grain. Heavy wood is harder than light wood. Seasoning tends to increase, and moisture to decrease, the hardness. Placing the annular rings in a vertical position appears to help the wood to resist indentation.

A tough wood is a wood that is strong and flexible and able to resist shocks and blows. Toughness is often measured in impact tests by finding the amount of work required to cause rupture. Wood which offers a high resistance to tension and longitudinal shear and which is capable of suffering a distortion of more than 3 per cent in tension and compression is usually tough.

209. Miscellaneous Properties of Timber.—The approximate chemical composition of all woods when dry is nearly uniform, and consists, by weight, of the following elements: 49 per cent carbon, 6 per cent hydrogen, 44 per cent oxygen, and 1 per cent ash. The weight per cubic foot for most dry woods varies between 25 and 60 lb. A few woods are lighter and some are heavier than these values. Moisture increases the weight per cubic foot very considerably.

The coefficient of linear expansion per degree Fahrenheit and for temperatures between 35 and 60 degrees Fahrenheit varies as follows: parallel to the fibers from 0.0000014 to 0.0000054; and perpendicular to the fibers, from 0.0000019 to 0.0000034.

210. Factors of Safety and Safe Working Loads for Timber.—The factors of safety (and safe working stresses) vary according to the kind of stress and kind of loading and also according to the judgment of various engineers. In designing, only the net section of the dressed timber should be considered.

The following factors of safety for variable loads are considered good practice: 10 for tension, 5 for compression with grain, 4 for compression across grain, 6 for extreme fiber stress in cross bending, 2 for modulus of elasticity in cross bending, and 4 for shear.

For steady loads, these factors of safety may be decreased 33 per cent (corresponding to an increase of 50 per cent in the unit

Properties of Timber—15 to 20 Per Cent Moisture

Kind of wood	Weight in pounds per cubic foot	Strength in pounds per square inch									
		Tension with grain. Ultimate	Compression				Cross bending			Shear	
			With grain			Across grain. Ultimate	Elastic limit	Ultimate	Modulus of elasticity	With grain. Ultimate	Across grain. Ultimate
			End bearing		Column under 15 diameter						
			Elastic limit	Ultimate							
Conifers											
White pine	24	7,000	4,000	5,500	3,500	700	3,000	4,000	1,000,000	400	2,000
Red (Norway) pine	31	8,000	3,500	5,000	4,000	800	3,500	5,000	1,100,000	500	3,500
Yellow pine (Long leaf)	38	12,000	5,000	7,000	5,000	1,400	5,000	7,000	1,500,000	600	5,000
Yellow pine (Short leaf)	32	9,000	4,500	6,000	4,500	1,000	4,500	6,000	1,200,000	400	4,000
Douglas (Oregon) fir	32	8,000	4,000	5,700	4,500	800	3,500	5,000	1,400,000	500	4,000
Spruce	25	8,000	4,000	6,000	4,000	700	3,000	4,000	1,200,000	400	3,000
Hemlock	25	6,000	3,500	5,200	4,000	600	2,500	3,500	900,000	350	2,500
Cedar	23	7,000	4,000	5,500	3,500	700	3,000	4,000	700,000	400	1,500
Tamarack	30	5,500	3,000	4,500	3,500	600	3,500	5,500	1,000,000	300	1,500
Cypress	29	6,000	3,500	5,000	4,000	700	3,500	5,000	900,000	500	3,000
Redwood (California)	24	7,000	3,500	5,000	4,000	600	3,000	4,500	700,000	400	2,000
Broad Leaved											
White oak	50	12,000	5,000	7,000	5,000	2,000	5,000	7,000	[illegible]	800	4,000
Hard maple	43	9,000	5,500	8,000	5,000	1,500	5,000	7,000	[illegible]	500	5,000
White ash	41	13,000	4,000	6,000	4,000	1,800	5,500	8,000	[illegible]	700	5,000
Black walnut	38	12,000	5,000	7,500	5,000	600	5,000	7,500	[illegible]	600	3,500
Elm	45	10,000	5,500	8,000	5,000	1,200	6,000	8,500	[illegible]	700	4,000
Hickory	49	14,000	5,500	8,000	5,000	2,900	7,000	9,500	[illegible]	1,000	6,000
Locust	45	15,000	5,800	8,500	5,000	2,000	6,500	9,000	[illegible]	1,000	6,000
Gum	36	16,000	5,000	7,000	5,000	1,400	5,500	8,300	[illegible]	700	5,500
Mahogany	35	9,000	4,000	6,000	4,500	1,000	6,000	9,000	[illegible]	600	4,000
Chestnut	41	8,500	3,500	5,000	4,000	900	3,500	5,000	[illegible]	600	2,000
Poplar (White wood)	30	7,000	3,000	4,500	3,000	500	3,500	5,000	[illegible]	500	4,000
Lignum-vitæ	77	10,500	6,000	8,500	5,500	2,200	6,000	9,000	[illegible]	800	4,000
Teak	52?	12,000	6,500	9,000	6,000	1,800	6,500	9,500	[illegible]	1,000	6,000
Catalpa	28	7,000	3,000	4,500	3,000	600	3,000	4,500	[illegible]	500	3,000
Eucalyptus	40?	10,000	4,500	7,000	5,000	1,500	5,000	7,500	1,000,000	800	4,500
Beech	40	8,000	3,500	5,000	3,500	1,200	3,500	5,500	[illegible]	600	4,000
Endogenous											
Palmetto	27	10,000	4,000	5,500	4,000	1,200	5,000	7,000	900,000		
Bamboo	32	13,000	6,000	8,500	6,000	1,800	8,000	11,000	1,400,000	1,000?	4,000?
Safety factor for working loads	..	10	.....	5	5	4	.....	6	2	4	4

working stresses). These factors of safety are too low for timbers containing large or loose knots.

For continuous heavy loading, loading causing a reversal in stress, or for apparently sound old timbers, the allowable unit stresses (safe working loads) should be about 80 per cent, or those obtained by the use of the factors of safety for variable loads.

211. Properties of **Timber.**—The preceding table gives the properties of timber containing from 15 to 20 per cent of moisture:

The above values are average values for dry commercial timber. Small specimens or specimens of exceptional quality may give test results 50 or 60 per cent higher. Specimens of poor quality (containing serious defects) or specimens from a weaker species of wood may give test results less than those given above. Also, green wood or wood containing a large amount of water may give results below those given above.

To obtain the working stress, divide the ultimate by a suitable factor of safety, depending upon the kind of load (see article on "Factors of Safety" and "Safe Working Loads for Timber").

E. SELECTION AND INSPECTION OF TIMBER

212. Selection **of** Timber.—When timber is selected for a special purpose it should be investigated and the kind chosen which appears to meet most fully the particular requirements of the case.

For framing timbers, woods should be selected that are plentiful and, consequently, cheap, and which can be obtained in large dimensions. Sometimes it is important to consider extra strength and durability.

For wood that is to be buried in the ground (either in whole or in part) or is to be used for piling, durability is the chief consideration, although the question of cost must be often considered.

For wood that is to be used in water (either in whole or in part) for piles, wharves, etc., durability and freedom from attack by borers are the most important considerations.

For outside finishing, ease of working and freedom from warping and checking are the most important requirements. The wood should be able to stand the wear from exposure (weathering).

For floors, the wearing qualities (and sometimes the appearance) are the chief considerations.

For interior finishing and decorating, the color and grain of the wood and its ability to take a polish and finish often decide the choice.

213. Inspection of Timber.—Timber is inspected to determine the quality of the stock and the dimensions of the pieces. All condemned pieces must be plainly marked with paint or a branding iron. Sometimes, in the case of large timbers, all accepted pieces are marked with a paint or a branding iron which is different from that used in marking the condemned pieces.

Strong and durable timber possesses the following characteristics:

It is obtained from the slowest growing trees, as indicated by the narrowness of the annular rings.

The best timber comes from the heart of the tree, and should include no sapwood.

Wood containing the least amount of resin or sap in its pores is the most durable.

The wood should be uniform in appearance, straight in fiber, free from large and dead knots, and free from all flaws, shakes, and blemishes.

Freshly cut sound timber smells sweet, and shows a firm and bright surface with a silky luster when it is planed.

The surface should never be woolly, and the wood should not clog the teeth of the saw.

In highly colored woods, darkness of color generally indicates strength and durability.

Sound timber, when lightly struck or scratched at one end, transmits the sound to the ear placed at the other end, even though the timbers are as long as 50 ft.

Sound timber is sonorous when struck, while decaying timber gives forth a dull sound.

A dull, chalky appearance and a disagreeable odor are signs of bad timber.

In the absence of the usual external signs, dry rot may be detected by boring test holes in the wood and then examining the appearance and odor of the wood dust.

Timber containing defects, such as knots, cheeks, cracks, pitch pockets, bark, excess sapwood, etc. not allowed by the specifications, should be rejected.

Timber should be measured (length, breadth, and thickness) to see if it is of the proper dimensions (see article on "Classification of Lumber").

F. PRESERVATION OF TIMBER

214. Preservation of Timber in General.—The life of timber can be prolonged somewhat by thorough seasoning, but a better way is to inject into the timber some substances such that, while they are not appreciably harmful to the timber, will act as poisons toward the fungi, borers, etc. that cause the decay of wood. The materials most commonly used are creosote (dead oil of coal tar), zinc chloride, copper sulphate, and bichloride of mercury (corrosive sublimate).

There are three general methods of injecting the "preservatives" into the timber. The first or pressure process (sometimes called the pressure-tank process) makes use of force pumps, air compressors, etc., to secure the pressure desired. Practically all of the special processes are pressure processes.

The non-pressure, or open-tank, process uses atmospheric pressure only. One method is to take thoroughly seasoned wood, immerse it from 1 to 6 hours in a bath of hot liquid, and then change it quickly to a cold bath. This change causes a contraction of the air and moisture in the timber and allows the entrance of the preservative. Another method is first to heat the timber in an oven and then suddenly immerse it in a cold bath of the preserving liquid.

The third general method consists of painting the timber with one or more coats of the preservative by means of a brush. It is known as the "brush" process. The painting, with coal tar, of timbers that are to be placed in the ground is a common example of this method.

The pressure process is thought to decrease slightly the strength and elasticity of the timber, but the other two general processes seem to have no bad effects, as the coating of the cell walls and fibers with a preservative should not injure them. However, a too concentrated solution of some of the preservatives might cause chemical dissociation of the wood. Preliminary steaming at too high pressures or for too long a time weakens the timber. The limits for pressure and time are about as follows: 40 lb. per square inch for 3 hours, 30 lb. per square inch for 4 hours, and 20 lb. per square inch for 5 hours.

215. Creosote Processes for Preservation of Timber.—Creosote has been found to be the best of all of the preservative coatings for timber under practically all conditions. It is especially effective against the Teredo and other sea worms and borers. Creosoting is not suitable for wood that is to be used for interior or decorative work. Any of the three general methods may be used for creosoting, but the pressure process is the best.

The open-tank process consists of immersing the timbers in a bath of creosote and allowing them to soak for a few days before they are removed and used. This process is not very satisfactory as the penetration of the oil is small.

Bethell's process of creosoting is probably the most important process in use. It is briefly as follows: The timber is placed in a large cylinder; steamed thoroughly for a few hours to evaporate the sap; then a pump removes the sap and steam and produces a partial vacuum; after which the cylinder is filled with creosote oil (distilled from coal tar) heated to about 150 degrees Fahrenheit and under a pressure of about 175 lb. per square inch. About 5 lb. of oil per cubic foot of timber are required. The length of time is 24 hours. Green timber requires from 12 to 18 lb. of oil per cubic foot of timber.

Seeley's process is a modification of Bethell's process. The timber is immersed in creosote oil at a temperature of 212 to 300 degrees Fahrenheit for a time sufficient to expel the moisture. The hot oil is then drawn off and replaced by a cold bath. The amount of oil absorbed is about 4 lb. per cubic foot of wood.

In the Breant process the timber is placed in a vertical cylinder and the liquid is let in almost to the top. A partial vacuum is then produced by an air pump, after which the valve is closed and the liquid forced in until a pressure of about 10 atmospheres is reached. The time required for impregnation is about 6 hours.

In the A. C. W. process the method is like the Bethell process except that an air pressure of 15 lb. per square inch is applied after the vacuum and maintained while the creosote is admitted so as to prevent unequal absorption during the filling. Then a pressure of 100 lb. per square inch is applied until the desired amount of penetration is reached, after which the creosote is withdrawn and air is forced in under a pressure of about 70 lb. per square inch.

In the boiling process (used principally for Douglas fir) the timber is placed in a cylinder containing creosote oil at a temperature a little above 212 degrees Fahrenheit and kept there for a period varying from a few hours to 2 days. Then a pressure of about 120 lb. per square inch is applied, the temperature is allowed to drop, and the preservative forced into the wood.

In the Ruping, or empty-cell, process the timber is air dried before being placed in the cylinder. Air is admitted at a pressure of 75 lb. per square inch and then the creosote is admitted at about 85 lb. per square inch, after which the pressure is increased up to about 225 lb. per square inch to force the oil into the timber. When the pressure is removed, the compressed air in the wood forces out most of the oil, leaving only a coating around the cell walls. This process gives a very high penetration of oil with but little absorption.

The Lowry process is the same as the Ruping process, except that no compressed air is used before admitting the creosote under pressure.

In the Kreodone process the timber is first sterilized by subjecting it to a dry heat of 240 degrees Fahrenheit for 8 hours. The remainder of the process is like the Bethell process, except that Kreodone (an oil derived from creosote) is used instead of the creosote. The absorption is about 12 lb. of oil per cubic foot of wood.

In the creo-resinate process a mixture of creosote and resin, containing from 50 to 70 per cent creosote, is used instead of the creosote. The process is similar to the Bethell process, except that a dry heat is used instead of the steam bath before the vacuum.

216. Zinc Chloride Pressure Processes for Preservation of Timber.—Burnett's process is the most important zinc chloride process. It is briefly as follows: The timber is placed in a closed metal cylinder; a 20-in. vacuum is produced; steam is added at a pressure of 25 lb. per square inch for about 4 hours; then the steam is removed and a second vacuum produced; a zinc chloride solution is added at a temperature of 150 degrees Fahrenheit and under a pressure of about 135 lb. per square inch. About 0.24 lb. of zinc per cubic foot of timber is required. The time taken is about 10 hours. The zinc solution contains about 43 per cent of zinc, 2 per cent of impurities, and 55 per cent of water.

In the Allardyce process a 2 or 3 per cent zinc chloride solution is first forced into the timber by a process like Burnett's process and then creosote is injected. The creosote remains on the outside, thus protecting the soluble zinc chloride in the interior. About 12 lb. of the zinc chloride solution and 3 lb. of creosote are required per cubic foot of timber.

In the card process the preserving liquid is a 3 to 5 per cent solution of zinc chloride containing about 15 or 20 per cent of creosote. The process is similar to that of Bethell's. As the two preservatives will not mix, a pump is used to keep them in a mechanical mixture.

The Wellhouse process consists of first steaming the timber in a cylinder for a few hours and then adding a solution of zinc chloride and glue under pressure, after which tannin is injected. The glue combines with the tannie acid in the wood and is precipitated while the wood retains the zinc. The tannin is added to precipitate the excess of glue.

217. Vulcanizing Process for Preserving Timber.—Vulcanizing consists of rendering the sap insoluble and undecomposable by the application of heat. This process tends to make the timber less combustible besides preserving it. The wood is placed in a closed vessel and under pressure to prevent the vaporization of the sap when it is heated. The heat is gradually applied and the pressure is gradually increased as the temperature rises. A temperature of 400 degrees Fahrenheit is sufficient to vulcanize ordinary woods. The time required varies from about 8 hours for soft woods to from 10 to 20 hours for hard woods.

218. Some Other Pressure Processes for Preserving Timber. Boucherie's (copper sulphate) process is the addition of copper sulphate under a pressure of 15 lb. per square inch and in the proportion of 1 lb. of copper to 10 gal. of water. This method is used very much in Germany. The wood should be green.

Kyan's process (sometimes called "Kyanizing") consists of impregnating the timber with bichloride of mercury (corrosive sublimate). The solution is in the proportion of 1 lb. of mercury bichloride to 50 lb. of water.

Payne's process consists of injecting sulphate of iron into the wood in a vacuum, followed by an injection of a solution of sulphate of lime or soda. This process also makes the wood incombustible.

Thilmany's process is the impregnation of the wood with zinc or copper sulphate. The method is about the same as Bethell's creosoting process, except that, instead of adding creosote, a solution of zinc sulphate (or copper sulphate) is added under a pressure of about 90 lb. per square inch. After the residue of sulphate is removed from the cylinder, a 1 per cent solution of barium chloride is added under pressure. Green wood is preferred for this process.

CHAPTER X

PIG IRON

A. DEFINITION OF PIG IRON AND ORES OF IRON

219. Definition of Pig Iron.—As pure iron is not found free in nature, it is necessary to reduce the ores of iron. The product obtained by the reduction of iron ores in a blast furnace is called pig iron. These iron ores usually consist of compounds of iron and oxygen with some other impurities. The important commercial ores contain from 25 to 70 per cent of iron.

220. Ores of Iron.—The ores of iron in the order of their importance are as follows:

Red Hematite (Fe_2O_3) varies from black to brick-red in color and contains about 70 per cent of iron when pure. The presence of impurities reduces this percentage to about 50 or 60 per cent. The specific gravity of the red hematite is about 5.3; its streak is always red. This ore often occurs as an earthy ore which is easily mined and handled. It is found on all of the continents and is the most important ore in the manufacture of iron.

Brown Hematite or Limonite ($Fe_2O_3 + nH_2O$) varies in color from a brownish-black to a yellowish-brown, usually occurs in massive form, and is softer and lighter than the red hematite. Limonite ore is a red hematite containing about 14.5 per cent of chemically combined water. Its streak is yellowish-black. Its specific gravity varies from 3.6 to 4.0. When pure, this ore contains about 60 per cent of iron, but the commercial ore usually contains from 40 to 50 per cent of iron. Limonite is found principally in the United States.

Magnetite (Fe_3O_4), the richest and hardest iron ore, is a hard black mineral occurring in a granular or massive form. It contains 72.4 per cent of iron when pure; has a specific gravity of about 5.2; has a black streak; is almost as magnetic as pure iron; and is often contaminated with oxides of silicon, phosphorus, and titanium. At present there is no economical method for separating the titanium oxide from the iron ore and, consequently, the ores having this impurity in them are practically worthless. Magnetite is found principally in Sweden and the United States.

Iron Carbonate ($FeCO_3$), often called siderite or spathic ore, is gray or brown in color and contains about 48 per cent of iron when pure. Ordinary ore, due to the presence of impurities, contains only from 30 to 40 per cent of iron. Its specific gravity varies from 3.7 to 3.9. If exposed to the weather for some time, this ore will change to limonite or red hematite. Iron carbonate is found in England, the United States, and other countries.

Pyrite (FeS_2) contains about 47 per cent of iron when pure. It is golden-yellow in color; has a greenish or brownish-black streak; and a specific gravity between 4.8 and 5.2. As this ore must be desulphurized before being used, very little of it is used at the present time.

221. Ore Mining.—The soft earthy ores that lie on the surface or near the surface of the ground are mined by steam shovels in an open cut. The soil is first removed from the top of the ore bed, and then the steam shovel loads the ore into ore cars which convey the ore to the blast furnace.

The hard ores must be drilled and blasted before they can be transported to the blast furnace. It costs more to mine the hard ores than the soft ones, but a certain proportion of hard ore is required for the successful operation of the blast furnace.

If the ore veins are underground, the mining must be carried on by underground methods. The expense depends upon the conditions present.

222. Preliminary Treatment of Iron Ores.—The rich soft ores (red hematite) require no preliminary treatment before they are fed to the blast furnace, but the other ores need to be treated.

Crushing to proper size is necessary for the hard ores. Jaw or gyratory crushers are usually used for this purpose.

Ores containing clay and dirt are washed to remove these impurities.

Calcination, or preliminary heating, is used to remove the water or carbon dioxide from the ores. Sometimes some of the gangue is oxidized at the same time. The heater used is a vertical furnace something like a mixed feed lime kiln.

If the ore contains sulphur, it may be roasted to remove this impurity. Only a moderate heat is required.

If the ore is magnetic, or has been made magnetic by calcination, a magnetic separator may be used to separate the ore from much of its gangue.

Ores may be divided into bessemer or non-bessemer ores,

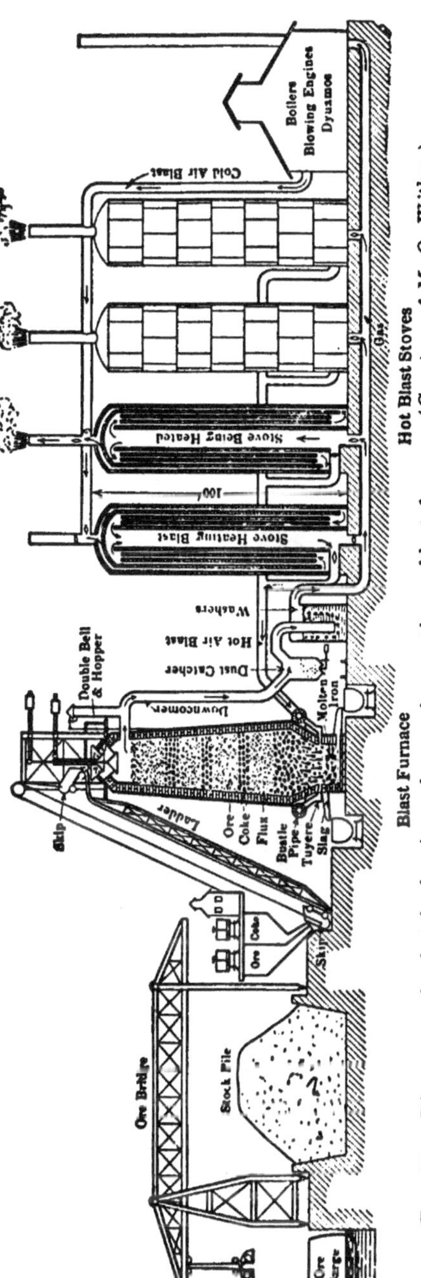

Fig. 78.—Diagrammatic sketch showing scheme of operating a blast furnace. (*Courtesy of M. O. Withey.*)

according to their phosphorus content. Ores in which the phosphorus content exceeds $1/1000$ part of the iron content are called non-bessemer ores.

B. MANUFACTURE OF PIG IRON

223. The Blast Furnace.—Practically all of the iron used today is first reduced from the ores to pig iron by means of the blast

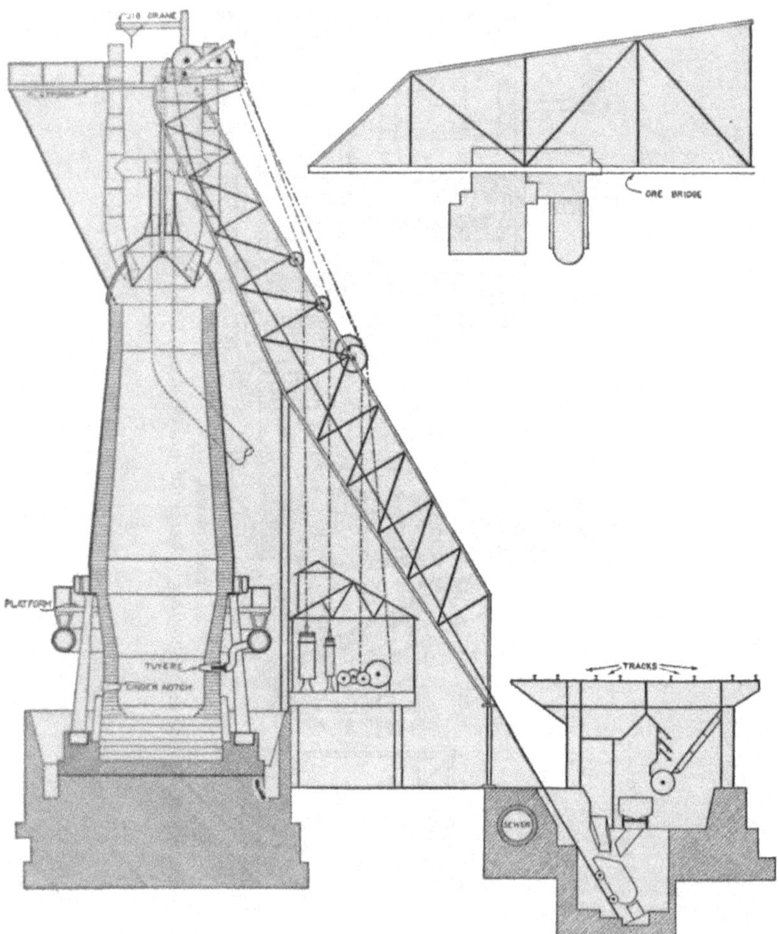

Fig. 79.—Blast furnace with charging mechanism. (*Illinois Steel Company*).

furnace. This furnace is a big circular steel shell, usually about 100 ft. high and 22 ft. in diameter, lined with silica firebrick. A furnace of this size has a capacity of about 450 tons of pig iron in a 24-hour day. About 900 tons of good ore are required to produce this 450 tons of iron.

The bottom part of the furnace, called the hearth or crucible, is about 7 or 8 ft. high, and its purpose is to hold the molten metal. At the lower part of the hearth there is a tap hole for drawing off the molten iron. About halfway up the hearth there is another hole, called the cinder notch, which is used to draw off the slag. When not in use, these holes are plugged with clay balls. Just below the top of the hearth there is a ring of nozzles or tuyeres

Fig. 80.—Blast furnace plant. (*Illinois Steel Co.*)

piercing the hearth linings. These tuyeres are connected on the outside to a large air pipe called the bustle pipe. The tuyeres admit the hot-air blast to the furnace.

The part of the furnace above the hearth is called the bosh. This is the hottest part of the furnace and it has to be cooled by some means such as spraying water against the outer surface or by the use of a water jacket full of running water. The bosh is about 12 ft. high.

The part of the furnace between the bosh and the bell and hoppers is called the stack. This stack is divided into three parts, called the lower inwall, middle inwall, and top. The height of the stack is about 50 or 60 ft.

The top of the blast furnace is provided with a double bell and hopper for charging the furnace with fuel, flux, and ore. The materials are first charged into the upper hopper. They fall into the lower hopper when the upper bell is lowered. Lowering the lower bell allows the materials in the lower hopper to spread in an even laver on the materials in the furnace.

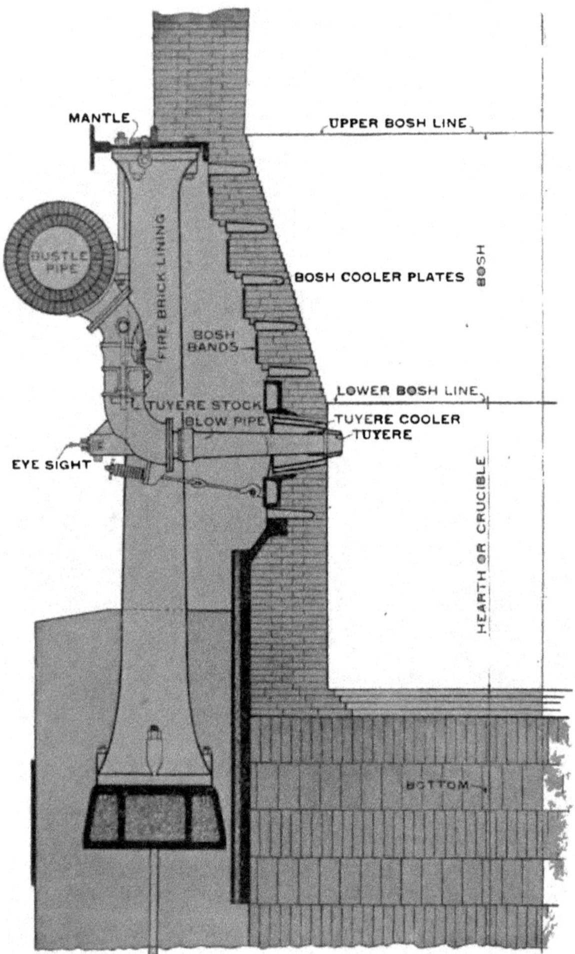

FIG. 81.—Sectional view of blast furnace showing details of hearth and bosh. (*Harbison-Walker Refractories Co.*)

The hot waste gases, generated during the operation of the furnace, are removed from the furnace by means of a pipe called the downtake. This pipe enters the furnace just below the lower bell. The downtake is provided with an extra pipe and valve (called the bleeder) which serve as a relief if the gas pressure gets too high. At the bottom of the downtake, there is a dust catcher which removes all of the solid particles in the gas passing through the downtake.

224. Accessories of the Blast Furnace.—The mechanical equipment used in connection with the blast furnace includes bins for holding the ore, fuel, and flux; charging equipment; blowing engines and auxiliaries; hot-blast stoves; and usually some apparatus for drying the blast.

The charging equipment consists of a double inclined track and skips or small cars which carry the materials from the bins to the top of the furnace. These cars are loaded by gravity at the bins, and they dump the materials automatically in the upper hopper.

The blowing engines are usually gas engines of about 2,000 horse power which will deliver from 45,000 to 60,000 cu. ft. of air per minute at a pressure of from 15 to 30 lb. per square inch. These engines use the washed waste gases from the furnace as fuel.

The hot-blast stoves are steel shells, about as high as the blast furnace, lined with firebrick and divided into a number of vertical compartments. There are usually four of these stoves for each furnace. The stoves are first heated by the escaping burned gases from the blast furnace and are then used to heat the air blast, when on its way to the furnace from the blowing engines, to a temperature of about 1,000 degrees Fahrenheit. Usually one stove is used to heat the blast while the others are being heated by the exhaust gases.

As ordinary air contains about 1 lb. of water per 1,000 cu. ft., it has been found economical first to dry the air before admitting it to the blowing engines. This drying is accomplished by first cooling the air down below its dew point by means of refrigerating machinery. This cooling causes a condensation of a large part of the moisture in the air. The drying of the air blast permits of a saving of about 15 per cent in the amount of fuel required in the blast furnace to reduce the ore.

225. The Fuel.—The fuel used in a blast furnace must provide the necessary heat and also act as a reducing agent. This fuel must be strong and porous and capable of producing an intense heat without leaving an ash which has a high content of phosphorus or sulphur. The fuels that are used in blast furnaces are coke, coal, and charcoal.

Coke is the solid residue obtained by the distillation of the volatile matter in certain grades of bituminous coal, called "coking coals." The weight of coke produced is about two-thirds of that of the coal used. At present, coke is the most used and the most satisfactory fuel for blast furnaces. Compared with anthracite coal and charcoal, coke is first in firmness, second in structure, and third in ash. The choice of fuel in any locality depends much upon the relative cost, charcoal usually being the most expensive.

Anthracite coal can be used successfully in the blast furnace without preparation. It tends to become broken into fine pieces upon heating, is less firm than coke, and has less ash than coke has but more than charcoal.

Bituminous coals are the least desirable to use as they soften and become broken up, when heated, besides containing more ash than any of the other fuels.

Charcoal has the least ash and the best structure of all the fuels, but is less firm than coke or anthracite. Charcoal is the carbonaceous residue remaining after wood has been heated and the volatile matter driven off. In making charcoal, air must not come in contact with the wood while it is being heated. Charcoal is fragile, light, and very porous.

226. The Flux.—A flux is a substance which will combine with the earthy parts of the iron ore and the ash of the fuel and produce a fusible slag which may be separated from the metallic iron. Acid gangues require basic fluxes, and *vice versa.* Practically all of the gangues are acid in character; hence, basic fluxes are nearly always required.

Limestone is the cheapest form of a basic flux, and it is used oftener than any other. The limestone should be very pure as acid impurities reduce the efficiency. Sometimes dolomitic magnesium limestones are used if the high-calcium limestones are hard to obtain. The presence of the magnesium appears to have no bad effects on the flux, and some think that magnesium improves the quality of the flux.

Quicklime could be used instead of limestone for a flux, but its advantages are more than offset by the increase in cost.

227. Operation of the Blast Furnace.—In reducing an iron ore, the blast furnace has the following five important functions to perform:

1. The deoxidation of the iron ore.

2. The carburization of the iron.
3. The melting of the iron.
4. The making of a fusible slag and then melting it.
5. The separation of the molten iron from the molten slag.

The charge of a blast furnace consists of about one-half ore, one-third fuel, and one-sixth flux deposited in the furnace in rotation and continually so as to keep the level of the materials

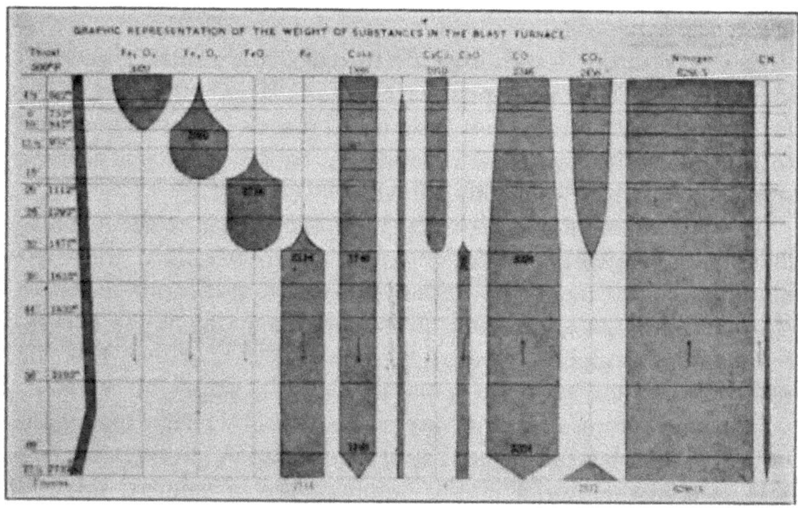

FIG. 82.—Chart showing changes occurring in a blast furnace charge. Figures represent lbs. per 2240 lbs. of pig-iron. Horizontal distances are proportional to weights. (*From Campbell.*)

about 7 or 8 ft. below the bottom of the lower bell and hopper. Different grades of ores are often mixed so as to secure the proper proportioning of the chemical elements.

The air blast enters the furnace through the tuyeres and rises through the various layers of the charge to the top. The oxygen of the blast combines with the carbon in the fuel and forms CO with the evolution of much heat, the temperature being about 3,500 degrees Fahrenheit at the top of the hearth. Passing through a layer of ore, the CO takes oxygen from the ore and forms CO_2 besides heating the ore. Reaching another layer of fuel, the CO_2 takes carbon from this fuel and forms CO which again changes to CO_2 when the next layer of ore is reached. This process is continued until the gas, consisting of N from the air blast and CO and CO_2, arrives at the top of the furnace, where it passes off through the downtake.

During the process of reduction of the iron ore, a thin layer of carbon is deposited upon the brickwork of the furnace. This layer protects the brick from rapid destruction due to the corrosive action of the slag.

The gas from the furnace, after passing off through the downtake and being cleaned, is used for heating the hot-blast stoves and for operating the blower engines and other engines used for supplying the power required to operate the machinery of the plant.

As the fresh ore descends in the furnace, it loses its oxygen and absorbs much heat and some carbon from the rising gases. The carbon makes the iron more fusible and it soon begins to melt and to run down through the fuel and slag to the hearth, absorbing some silicon and sulphur on the way.

The limestone flux loses CO_2 near the center of the furnace, leaving the infusible calcium oxide which combines with the alumina, silica, and earthy bases in the ore and forms a fusible slag which floats on the top of the molten metal in the hearth. Some of the sulphur present combines with the lime, and the resulting calcium sulphide is dissolved in the slag.

Whenever necessary, the slag is drawn off through the cinder notch. This slag can be used as an aggregate in making concrete and as a raw material in the manufacture of cement, mineral wool, and paint.

When sufficient molten iron has collected in the hearth, it is drawn off through the tap hole, skimmed of its slag, and is either cast into pigs or conveyed in a molten condition to a steel plant.

228. Use of the Electric Furnace in Reducing Iron Ores.—Electric furnaces have been successfully used for the commercial reduction of iron ores in localities where electricity is cheap and the proper fuels are expensive. As only about two-ninths or one-third as much carbon is required with the electric furnace as with the ordinary blast furnace, there is a great saving of fuel. However, the cost of electrical heat is usually more than the cost of heat produced by the combustion of fuel.

The electric furnace that has been used the most is of the "resistance" type which uses electrodes that project into the charge or bath.

A very high quality of pig iron is produced by the electric furnace, which is an important consideration in the making of very high grade steel.

229. Making the Pigs.—A "pig" of cast iron is generally semi-cylindrical in form, about 5 in. in width, 36 in. in length, and 100 lb. in weight.

Formerly, the pig iron was cast in sand beds which consisted of a series of depressions molded in a bed of silica sand. The molten iron was withdrawn from the blast furnace through the tap hole, the slag skimmed off, and the iron allowed to flow through channels to the pig molds. When the iron had solidified, the iron in the pig molds was broken away from that in the channels, and then the iron in the channels was broken into suitable lengths by the use of sledge hammers. The channels are often called "sows." At the present time, most of the pig iron is cast in a pig-molding machine. Such a machine consists essentially of a number of steel molds fastened to an endless chain. The molten iron is carried from the blast furnace in a ladle and poured into the molds as they are carried past by the endless chain. The molds are carried through a water bath which solidifies the pigs, and then the pigs are dumped from the molds into a pile or into a car for shipment. As the molds return toward the charging ladle, they are sprayed with lime water to prevent the molten iron from sticking to them.

C. CLASSIFICATION AND USES OF PIG IRON

230. Classification of Pig Iron.—Following are four methods of classifying pig iron:

1. Pig iron may be classified according to the method of manufacture into coke pig, charcoal pig, and anthracite pig.

2. The second method is based on the chemical composition. There are four divisions: silicon pig which has a high silicon content; low phosphorus pig; special low phosphorus pig; and special pigs such as spiegeleisen, ferromanganese, ferrochrome, etc.

3. For commercial purposes, pig iron may be graded by the color, hardness, and character of the fracture into nine divisions: No. 1 Foundry, No. 2 Foundry, No. 3 Foundry, No. 1 Soft, No. 2 Soft, Silver Gray, Gray Forge, Mottled, and White. No. 1 Foundry is the most open grained, soft, gray iron and is largely granular in fracture. The other grades gradually pass from the soft gray iron into the hard white iron. Formerly, this classifica-

tion was much used but at the present time most of the pig iron is purchased by chemical analysis.

4. This is the best classification and is according to the proposed use of the pig iron. There are five divisions:

1. *Bessemer Pig.*—For acid bessemer or acid-open hearth processes of making steel.
2. *Basic Pig.*—For basic bessemer or basic open hearth-processes of making steel.
3. *Malleable Pig.*—For malleable cast iron.
4. *Foundry Pig.*—For gray cast iron.
5. *Forge Pig.*—An inferior foundry pig used for the manufacture of wrought iron.

The purchase of these grades is practically based on the chemical analysis. The following table gives the specified limits. In general, the carbon content varies from 2 to 4 per cent and the manganese content from 0.10 to 1.75 per cent.

CHEMICAL CONTENTS OF SOME PIG IRON

Kind of iron	Silicon, per cent	Sulphur, per cent	Phosphorus, per cent	Remarks
Bessemer pig	1.00 to 2.00	0.05 or less	Not over 0.10	Graphitic carbon,
Basic pig	Under 1.00	Under 0.05	Not specified	combined carbon,
Malleable pig	0.75 to 2.00	Not over 0.05	Not over 0.20	and manganese
Foundry pig	1.50 to 3.00	Not over 0.05	0.50 to 1.00	may be specified.
Forge pig	Under 1.50	Under 0.10	Under 1.00	

231. Uses of Pig Iron.—Pig iron, as such, has no structural uses but it is used in making commercial cast iron, malleable iron, wrought iron, and steel. More than 60 per cent of pig iron is used in the manufacture of steel by the bessemer and open-hearth processes. Approximately 3 per cent is used for making wrought iron and another 3 per cent for making malleable cast iron, while about 20 per cent is used in the manufacture of gray cast iron.

At the present time the United States of America is the largest producer and user of pig iron, producing and using about one-half of the world's supply of iron ore, pig iron, and iron products. This country produces annually about 70,000,000 long tons of iron ore from which about 35,000,000 long tons of pig iron are made.

CHAPTER XI

CAST IRON

A. DEFINITIONS AND GENERAL CLASSIFICATIONS

232. Definitions of Cast Iron.—The following are two definitions of cast iron:

Cast iron is iron containing so much carbon, or its equivalent, that it is not usefully malleable at any temperature.

Cast iron is a saturated solution of carbon in iron, the amount of carbon ordinarily varying from 2.0 to 4.0 per cent, depending upon the amounts of silicon, sulphur, phosphorus, and manganese present in the solution.

233. General Classification of Iron and Steel.—This general classification has been taken from the American Civil Engineers' Pocket Book.

Classification of Iron and Steel

Material	Per cent of carbon	Specific gravity	Properties	
Cast iron......	2.00 to 5.00	Average 7.2	Not malleable	Not temperable
Steel..........	0.10 to 1.50	Average 7.8	Malleable	Temperable
Wrought iron..	0.05 to 0.30	Average 7.7	Malleable	Not temperable

It is to be noted that the percentage of carbon alone is not enough to distinguish steel from wrought iron, as the process of manufacture and the presence of different chemical elements both have some influence on the properties. Also, only hard steels can be tempered, while soft steels resemble wrought iron in their properties.

234. Howe's Classification of Iron and Steel.—

I. *Carbon Class.*—Properties chiefly determined by the carbon content.

A. Weld Metal. Aggregated from a pasty mass without later fusion.

1. Wrought Iron. Carbon, 0.20 per cent or less.

2. Blister Steel. Carbon, 0.20 to 2.20 per cent.

B. Cast or Ingot Metal. Metal that is initially cast and is not aggregated.

1. Steel. Malleable when cast. Carbon, 0.20 to 2.20 per cent.

Low-carbon, medium, and high-carbon steel, depending upon the carbon content. Not dependent on the method of manufacture.

2. Cast Iron. Not malleable. Carbon 2.20 per cent or more.

Gray Cast Iron. Carbon segregated from the iron in form of graphite.

White Cast Iron. Carbon combined with the iron.

Mottled Cast Iron. Mixture of gray and white cast irons.

3. Malleable Cast Iron. Cast as white cast iron and then made malleable.

II. *Alloy Class.*—Properties chiefly determined by elements other than carbon.

A. Special or Alloy Steels. Used for special purposes and usually as a finished product.

B. Ferro Alloys. Alloys of iron used principally to introduce certain elements into steels.

B. MAKING THE MOLTEN CAST IRON

235. The Materials.—Cast iron is usually made by remelting pig iron in a cupola or air furnace and then casting it in its final form in molds. Sometimes the castings are made from molten pig iron taken directly from the blast furnace.

The materials used are pig iron, fuel, and flux. Often the pig iron from several blast furnaces is mixed so as to secure the proper proportioning of the chemical elements in the charge. Sometimes part of the pig iron is replaced by scrap iron (discarded castings), old or broken castings, sprues, etc.

Coke is almost always used for fuel, though in a few instances mixtures of coke and anthracite coal have been used instead. The amount of fuel required varies according to the castings and the furnaces. The cupola requires about 400 or 500 lb. of coke per ton of iron for small thin castings, and about 200 lb. of fuel for the larger castings. Small thin castings require a hotter metal in order to secure good results. The air furnace requires about twice as much fuel per ton of iron as the cupola does.

A little limestone, equal to about 1 per cent of the iron, is added as a flux. Its purpose is practically the same here as in

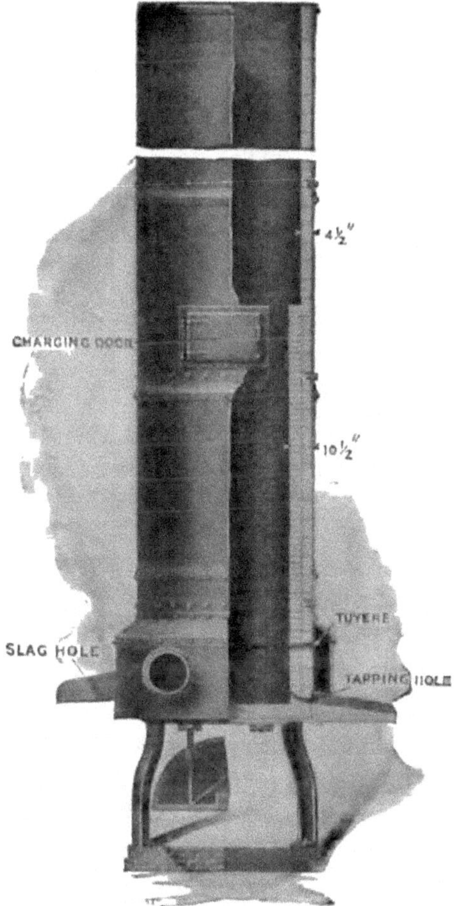

Fig. 83.—Section of typical cupola.

the reduction of the iron ores; namely, to combine with the gangue in the pig iron and the ash of the fuel and to produce a fusible slag which will float on the surface of the molten iron.

236. The Cupola.—The cupola, which is generally used for remelting the pig and scrap iron, is somewhat like a small blast furnace. It consists of a steel shell lined with firebrick and provided with a row or two of tuyeres near the base, through which the air blast enters at a pressure of about 1 lb. per square inch. The hearth, with a cinder notch for slag and a tap hole for the molten iron, is at the bottom of the furnace. The charging door is located on the side and a little more than halfway

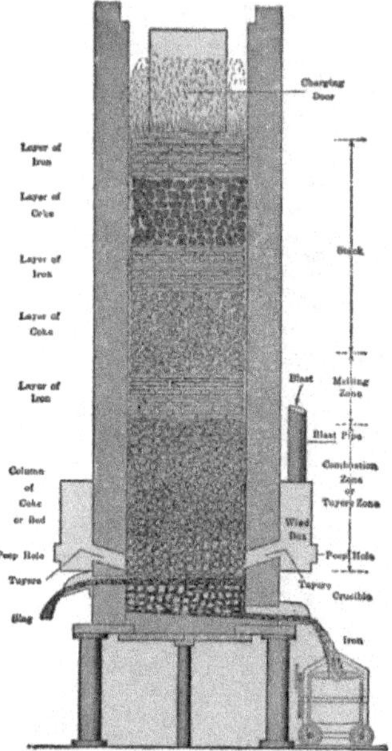

Fig. 84.—Zones in cupola. (*Stoughton.*)

toward the top of the stack. The stack is open at the top to allow the escape of the burned gases.

A cupola may be divided into four zones: (1) The bottom part or hearth which contains the molten iron and slag; (2) the tuyere or combustion zone where the air blast enters the furnace and the greatest heat is developed; (3) the melting zone, which is located just above the combustion zone, where the iron starts to melt; and (4) the stack zone which extends from the melting zone to the charging door.

The operation of the cupola is practically the same as that of a small blast furnace, except that there is very little deoxidation, the amount of fuel used is less, the pressure of the air blast is less, and no use is made of the escaping gases. About 100 cu. ft. of air is required per pound of coke. The cupola is operated almost continually. It is used for making practically all of the gray cast iron.

237. The Air Furnace.—The air furnace consists essentially of a hearth with a sloping roof, a grate or firebox at one end, and a stack with suitable flue and draft arrangements at the other end. The walls of the furnace are of heavy brick masonry reinforced with cast-iron plates fastened together with wrought-iron bolts. The hearth bottom is a mixture of sand and fireclay. Openings are provided for charging the furnace and for removing

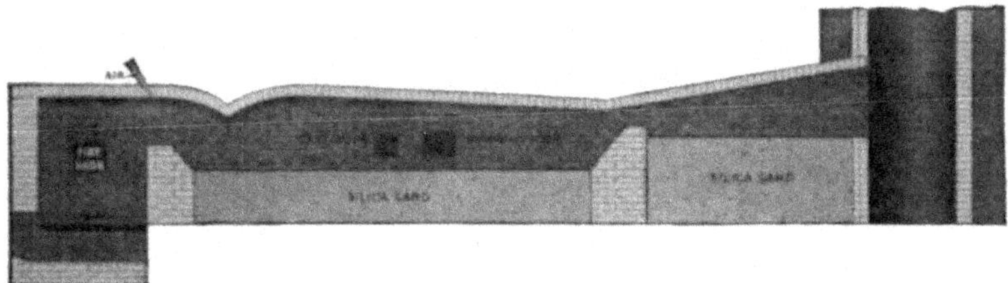

Fig. 85.—Longitudinal section of an air furnace.

the slag and molten iron. This furnace is somewhat like the furnace used in the wet puddling process of making wrought iron.

The air furnace is used for making white cast iron, malleable castings, and special cast irons of particular compositions.

In operating the air furnace, the fire is first started and the furnace heated, and then the pig iron, scrap iron, flux, etc. are placed on the hearth. The temperature is controlled by dampers in the firebox and stack. The furnace man rabbles the bath and skims off the slag from time to time. The length of time required is greater than in the case of the cupola.

C. FOUNDRY WORK

238. Definition of Founding.—Founding is the process of making iron castings by pouring the molten iron into molds of the desired shape and size which the metal assumes upon becoming cold.

In founding, the following things are important:

1. The selection and preparation of the iron used so that it will have the correct proportions of certain chemical elements required to make good castings.
2. Patterns and cores which are properly designed and made.
3. The selection of the proper materials for, and the preparation of, the hollow molds which receive the molten metal.

239. Patterns and Cores.—The patterns are usually made of a thoroughly seasoned wood, white pine or mahogany, and shellacked to keep the wood from being affected by moisture. Sometimes they are made of metal or plaster of Paris.

Patterns must be made so that they can be removed from the mold. They should have no sharp corners as these cause internal stresses when the casting shrinks in cooling. Fillets may be placed in sharp corners to round them. As cast iron shrinks in cooling, the patterns must be made a little larger to allow for this shrinkage. Pure iron will shrink 3/10 in. for each foot in length, while impure iron will shrink to a lesser extent, depending upon the amount and kind of impurities present. The allowance made for shrinkage varies from 1/32 in. per foot in large and bulky castings to 1/8 in. per foot for bars. Often the patterns are made to come apart in two or more places so as to aid in making the molds. The surfaces of patterns should be tapered so that they can be easily withdrawn from the mold. The making of the patterns is of great importance in foundry work and should be done only by workmen skilled in this work.

The cores, which are used to form the hollow spaces in the castings, are made of damp sand or clay for small molds and are constructed of built-up brickwork plastered with loam for the large castings, such as for large water pipes, etc.

240. Molds.—Molding sand consists mostly of silica, from 90 to 95 per cent, which makes the sand refractory; from 4 to 8 per cent of alumina and magnesia which furnish cohesion and plasticity; about 1.5 per cent or less of iron oxide; and about 0.5 per cent of lime. Very little iron oxide and lime are desirable as they make the sand less refractory. Sand that is very fine compacts too much, thus preventing the gases from escaping. This confined gas causes the formation of blow holes. Coarse sand lacks cohesion and makes inferior castings.

Green sand molds are the most commonly used molds. They are made of molding sand which is first uniformly dampened and then lightly rammed around the pattern to obtain the desired shape. The molds are usually made in open frames of iron or wood called core boxes or flasks. The core boxes often consist of two or more parts for convenience in molding. Besides the space for the casting, there must be an opening leading from the outside of the mold to the casting space to permit of the pouring of the molten iron. There should also be a skimmer to remove the slag, and a dirt riser between the pouring gate and the casting. At the other end of the casting, there should be a feeding gate. These openings are usually made by inserting plugs in the sand when the mold is made, and then removing the plugs when it is time to pour the iron. Small vent holes must be made in the mold to allow of the escape of the gases when the iron is poured.

For dry sand molds a loamy sand is used which is first roughly molded into shape and then dried in an oven and finished with a

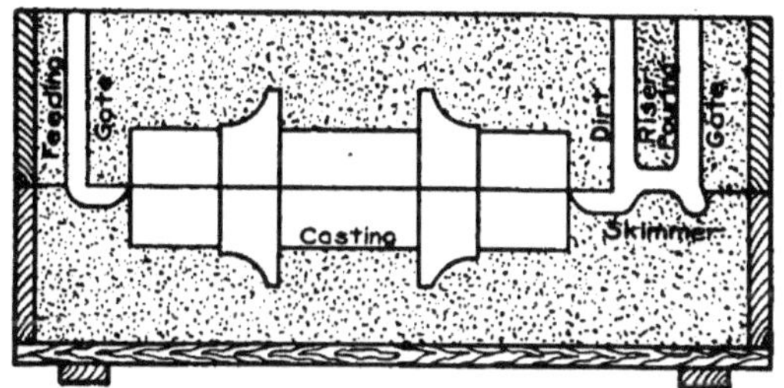

Fig. 86.—Section of a green sand mold.

tool. These molds give better castings than the green sand molds do and they require no patterns, but they are more expensive.

Loam molds are generally used for large castings. These molds are built up of brickwork, and the surfaces are finished by plastering with loam. The whole must be thoroughly dried before using. Loam molds are the most expensive molds.

Chills are metal molds used for certain castings, or are simply pieces of metal placed in other molds. This metal causes the molten iron that comes in contact with it to chill, or cool rapidly, and form a hard surface. In order to prevent an explosion, when the hot metal comes in contact with the cold, it is necessary to heat the chills to a temperature of about 350 degrees Fahrenheit before pouring the iron. These molds are used in making cast car wheels, chilled rolls, etc.

241. Pouring and Cleaning the Castings.—When the mold is ready, molten pig iron is drawn from the cupola into a ladle and conveyed to the mold. The molder holds the ladle as close as he can to the pouring gate and then pours the molten iron into the mold. If it is a top pouring ladle, a bar is used to keep back the slag. The molder stops pouring when the metal appears at the top of the riser. Each mold must be filled in one operation, or planes of separation will be formed in the casting. When the castings have become solid, they are removed from the molds and allowed to cool.

After the castings have cooled, they usually have to be cleaned of the sand that sticks to them. For small castings, a common method of cleaning is to place them in a tumbling barrel with some pieces of very hard iron. The barrel is rotated slowly and the tumbling about of the castings knocks the sand and scale off of their surfaces.

A better method of cleaning is to pickle the castings by immersing them in a dilute solution of sulphuric or muriatic acid. The time required is about 12 hours in a 15 per cent solution. When the castings are removed from the pickling solution, they must be carefully washed in water to remove all traces of the acid.

The sand blast is a very convenient way of cleaning the castings, especially those which are large or which have irregular surfaces.

The irregularities left by breaking off the gates, risers, etc. may be removed with a file, an emery wheel, or a cold chisel and hammer. Sometimes a pneumatic chipping tool can be economically used.

242. Defects in Castings.—The most common defects in castings are:

1. *Blow Holes.*—These are caused by the formation of steam when the hot metal comes in contact with the damp sand of the molds.

2. *Sand Holes and Rough Surfaces.*—These are caused by the failure or breaking down of the molds in places.

3. *Cracks.*—These are caused by unequal shrinkage in different parts of the casting when it cools.

4. *Cold Shorts or Seams.*—These are caused by cooling the iron so quickly that it does not completely fill the molds.

D. CONSTITUTION AND COMPOSITION OF CAST IRON

243. Constitution and Composition of Cast Iron in General.—Cast iron usually contains from 90 to 96 per cent of iron; 2½ to 4½ per cent of carbon; 0.5 to 4.0 per cent of silicon; and small percentages of manganese, phosphorus, sulphur, and other chemical elements.

The most important element in cast iron, besides the iron, is carbon. It may be present as free carbon (graphite) or as combined carbon. Carbon combines with iron and forms cementite, Fe_3C. This cementite may combine with free iron (ferrite) and form pearlite, which consists of about 6 parts of ferrite to 1 of cementite. Thus a cast iron may be composed of ferrite, graphite, cementite, and pearlite.

Ferrite, or free iron, is soft, weak, and tough. Graphite is weak. Cementite contains 6.67 per cent of carbon combined with iron, that is, 1 part of carbon and 15 parts of iron. It has great strength, is very brittle, and is harder than the hardest steel. Pearlite is more ductile, less hard and less strong than cementite, but is stronger and harder than ferrite.

If a section of ordinary cast iron is examined under the microscope, flakes of graphite will be found enmeshed in a matrix composed of ferrite, cementite, or pearlite, or combinations of these. The proportions of the four materials vary considerably in different cast irons, some cast irons containing practically no free carbon and others containing practically no combined carbon.

When the cast iron is in a molten state, the carbon is in solution with the iron. Silicon and aluminum tend to decrease the amount of carbon that can be held in solution, while manganese and chromium increase the solubility.

Slow cooling in the solidifying of cast iron allows the carbon to separate out and appear in the form of flakes (graphite). The presence of aluminum and silicon aid in this separation of the carbon and iron during the solidifying. Slow cooling tends to produce what is known as gray cast iron.

If cast iron is cooled rapidly from the molten state, the carbon tends to stay combined with the iron (because it does not have time to separate out) and produce what is known as white cast iron. Chromium and manganese, when present, help to keep the carbon in the combined form.

244. Effect of Carbon in Cast Iron.—Carbon has more effect on the properties of cast iron than any other element present, excepting the iron itself. Carbon usually varies in amount from 3 to 4 per cent. The properties of cast iron depend to a large extent on the amount of carbon present and also on the form that it is in, *i.e.*, combined or free.

Cast iron is usually classified according to the different form in which the carbon occurs. There are three different classes as follows:

1. White Cast Iron—where the carbon is combined with the iron.

2. Gray Cast Iron—where the carbon has separated from the iron and is present in the form of graphite.

3. Mottled Cast Iron—a mixture of gray and white cast iron.

The amount of combined carbon has a great influence on the properties of cast iron. This influence may be summarized briefly as follows:

As the ratio of combined carbon to total carbon increases from zero to one: The name of the matrix changes from low-carbon steel to medium carbon steel, then to high-carbon steel, and then to white cast iron.

The name of the cast iron as a whole changes from open gray or very graphitic cast iron to close gray cast iron, then to mottled cast iron, and then to white cast iron. The percentage of ferrite decreases while the percentage of cementite increases.

The strength increases for a while (until the percentage of combined carbon is about 1.2 per cent) and then decreases.

The hardness and brittleness increase, while the ductility decreases.

The ability to resist shock (toughness) decreases rapidly.

The ease of machining decreases.

White cast iron consists essentially of cementite and pearlite and is harder and more brittle than gray cast iron.

In gray cast iron, the amount of graphite varies from 2 to 4 per cent, while the amount of the combined carbon is less than 1½ per cent. Gray cast iron is composed of a mixture of ferrite, graphite, and cementite. This iron is softer, tougher, weaker, and less brittle than white cast iron.

Mottled cast iron is a mixture of particles of white and gray cast irons and contains ferrite, graphite, cementite, and pearlite in various proportions. Its properties depend upon the relative amounts of gray and white cast iron.

245. Effect of Silicon, Sulphur, Phosphorus, and Manganese on Cast Iron.—Ordinary cast iron contains, besides iron and carbon, four other chemical elements that are of importance, namely silicon, sulphur, phosphorus, and manganese. Cast iron also often contains very small percentages of other chemical elements such as aluminum, oxygen, nitrogen, copper, nickel, tin, chromium, etc. These latter elements are not present in large enough quantities in ordinary cast iron to cause any noticeable effect upon its properties. .

Silicon.—The amount of this element present usually varies from 0.5 to 4.0 per cent, and it aids in determining the suitability of the iron for various purposes. A little silicon, from 0.8 to 1.8 per cent, makes the iron soft and tough and gives the best strength results. More or less silicon tends to make the iron brittle and hard. About 3 per cent makes the carbon separate out in flake form (graphite). Silicon aids in foundry work by tending to increase the fluidity, eliminate the blow holes, and decrease the shrinkage when properly used. It reduces the chill in casting.

Sulphur.—This element helps to keep the carbon in the combined form and tends to make the iron hard, brittle, and weak. It also causes "red shortness;" *i.e.*, makes the iron very brittle at a red heat. Such iron is not good for steel manufacture. In foundry work, sulphur reduces the fluidity and increases the chill. Good cast iron rarely contains more than 0.15 per cent of sulphur.

Phosphorus.—The fusibility and fluidity of the iron are increased by from 2 to 5 per cent of phosphorus, thus helping in the making of fine castings in the molds. From 1.0 to 1.5 per cent of phosphorus is often used for fluidity and softness, but more than 1.5 per cent tends to make the iron brittle and hard. For the best strength results, not over 0.55 per cent of phosphorus should be present. Phosphorus tends to reduce the shrinkage and chill in castings. If the iron is to be used for steel making by the acid bessemer or open-hearth processes, the phosphorus content should be less than 0.07 per cent.

Manganese.—The amount of this element present may vary from 0 to 80 per cent, but rarely exceeds 2 per cent in ordinary castings. Iron that is to be used for steel making should have some manganese present, as the manganese tends to prevent the absorption of sulphur in remelting and also helps to neutralize the silicon besides making the steel more workable. Less than 1.0 per cent of manganese has practically no effect on the iron, while about 1.5 per cent makes the iron fine grained and hard to tool. Foundry iron usually contains less than 1.0 per cent of manganese. Manganese increases the shrinkage, decreases the magnetism, and increases the solubility of carbon in iron. Speigeleisen is iron containing from 10 to 50 per cent of manganese. It is capable of taking a high polish and is very hard, resisting cutting by hard cast steel tools. If the iron contains more than 50 per cent of manganese, it is called ferromanganese.

246. Effect of Some Other Chemical Elements on Cast Iron.—Many other elements are often found to some extent in cast iron in very small quantities, and these elements may have some effect on the properties of the iron.

Copper.—From 0.1 to 1.0 per cent closes the grain of cast iron, but does not appreciably cause brittleness. It makes the iron unsuitable for making malleable iron.

Aluminum.—From 0.2 to 1.0 per cent (added to the ladle in the form of a FeAl alloy) increases the softness and strength of white iron; added to gray iron, it softens and weakens it. About 0.1 per cent of aluminum has the same effect as 1.5 per cent of silicon. Aluminum is undesirable.

Tin.—Increases hardness and fusibility and decreases malleability besides making the iron unfit for conversion into malleable cast iron.

Vanadium.—Very small quantities increase softness and ductility. As much as 0.15 per cent added to the ladle in the form of a ground FeVa alloy greatly increases the strength of cast iron. Vanadium also acts as a deoxidizer and as an alloying material.

Titanium.—Increases the strength, when added in small amounts, such as a 2 per cent or a 3 per cent TiFe alloy containing about 10 per cent of titanium.

E. PHYSICAL AND MECHANICAL PROPERTIES AND USES OF CAST IRON

247. Strength of Cast Iron in General.—In general, large castings are not so strong as small ones. The shape of a casting also affects the strength; sharp reentering angles cause planes of weakness in cooling, while curved surfaces are not so weakened. The design of the castings is of importance. The methods of founding also greatly influence the strength. For example, if a hollow column is east in a horizontal position, the slag and foreign materials will tend to collect on the upper side and weaken that part of the column. Casting the column in a vertical position will cause the strength of all sides to be the same. As noted in previous articles, the presence of different chemicals, especially carbon and its combinations, have a great influence on the strength and properties of cast iron.

248. Tensile Strength of Cast Iron.—The combined carbon (for a total carbon content of about 4 per cent) in cast iron should be between 0.6 and 1.2 per cent to obtain the maximum tensile strength.

The tensile strength of cast iron varies from 10,000 to 45,000 lb. per square inch, but for ordinary castings it may be taken

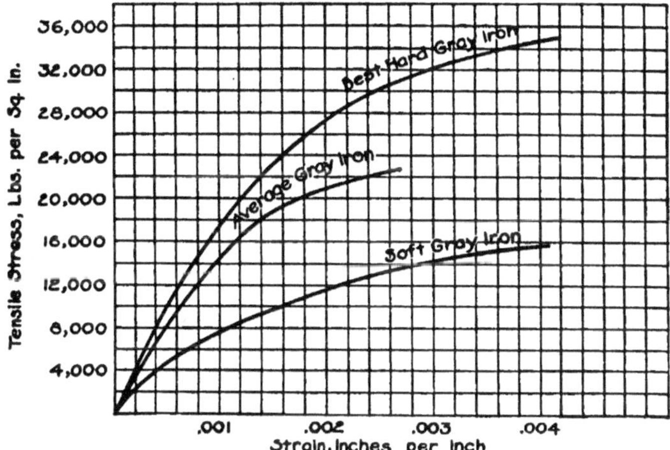

FIG. 87.—Stress-strain diagrams for cast irons in tension.

between 15,000 and 30,000 lb. per square inch. The minimum specification requirements for gray iron castings are: 18,000 lb. per square inch for lightweight castings; 21,000 lb. per square inch for medium castings; and 24,000 lb. per square inch for heavy castings.

The stress-strain diagram for tension is a curved line which shows no well-defined elastic limit; but, if the elastic limit is considered as the unit stress at which permanent set takes place, cast iron may be said to have an elastic limit which varies from 30 to 60 per cent of the ultimate, according to the grade of the iron.

The modulus of elasticity in tension is a variable quantity, depending upon the kind of iron and the unit stress at which it is computed. It will average about 14,000,000 lb. per square inch with probable variations of 25 per cent or more.

The percentage of elongation is very small, rarely exceeding 3 or 4 per cent for any grade of cast iron.

The reduction of area is so small as to be inappreciable.

The fracture of cast iron in tension is square across; that of gray cast iron being gray in color, highly crystalline in appearance, and showing flakes of free graphite, while the fracture of white cast

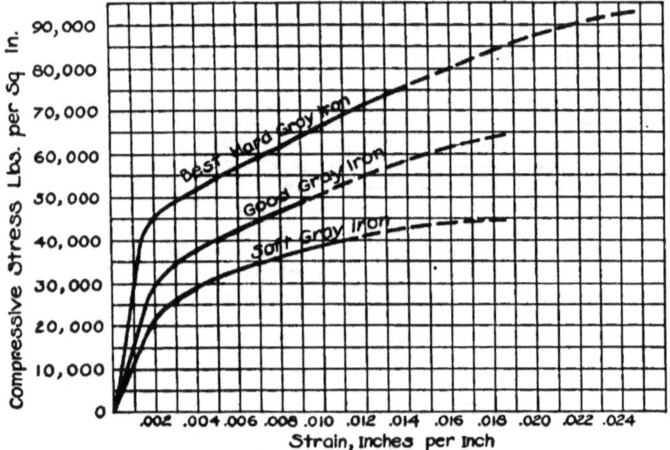

FIG. 88.—Stress-strain diagrams for cast irons in compression.

iron has a white metallic color and a finely crystalline appearance.

249. Compressive Strength of Cast Iron.—The compressive strength of cast iron depends upon about the same factors as the tensile strength. For the best results, the percentage of combined carbon should be a little more than that for tension, say from 1.0 to 1.2 per cent for a total carbon content of about 4.0 per cent.

In compression, the strength of cast iron may be taken from 60,000 to 100,000 lb. per square inch, though tests have shown variations from 45,000 to 200,000 lb. per square inch, depending upon the kind of iron, structure, composition, size, etc.

The stress-strain diagram for compression shows a fairly well defined yield point, varying from 35 to 60 per cent of the ultimate, depending on the grade of the iron.

The modulus of elasticity in compression varies, according to the kind and grade of the iron and the unit stress at which it is computed, from 10,000,000 to 25,000,000 lb. per square inch with an average of about 14,000,000 lb. per square inch.

When cast iron is stressed to failure in compression, the failure is usually shear along a plane making approximately 55 degrees with the line of loading.

250. Transverse Strength of Cast Iron.—At the present time the cross-bending strength of cast iron is the most important criterion of its quality, the tensile strength probably ranking second. The transverse strength is usually expressed by the term "modulus of rupture" and is computed by the formula $S = Mv/I$.

An average value of the transverse strength of cast iron is about 35,000 or 40,000 lb. per square inch, though tests have shown a variation of from 10,000 to 65,000 lb. per square inch, depending on the grade of iron, the structure, method of founding, length of span, etc. The percentage of combined carbon for the greatest strength should be between the values for the highest strength in compression and tension.

The minimum requirements for strength in transverse tests on the "arbitration test bar" over a 12-in. span are as follows:

Casting	Center Load, Pounds	Modulus of Rupture, Pounds per Square Inch
Light castings	2,500	39,000
Medium castings	2,900	45,000
Heavy castings	3,300	52,000

The deflection at the center shall not be less than 0.10 in.

The "arbitration test bar" is a bar 15 in. long and 1¼ in. in diameter, cast vertically in a green sand mold that is cold and thoroughly dry when the iron is poured. Two bars should be cast from each heat.

The "modulus of shock resistance" of cast iron is equal to the product of one-half the center load and the deflection divided by the volume of the specimen between the supports. It is approximately the same as the energy of rupture.

The failure of cast iron in cross bending is a failure by tension on the tension side of the beam.

The modulus of elasticity of cast iron in cross bending is about the same as that in tension and compression, averaging about 14,000,000 lb. per square inch with large variations depending upon the grade and structure of the iron, etc.

The modulus of elasticity in cross bending may be considered as a measure of the stiffness of the cast iron, or its ability to resist transverse loads without bending much.

251. Miscellaneous Properties of Cast Iron.—The computed ultimate resistance to shear and torsion varies from 20,000 to 35,000 lb. per square inch for cast iron.

The fusibility of cast iron depends upon the percentage of carbon and some of the other elements. An average value is 2,500 degrees Fahrenheit.

The coefficient of expansion is about 0.0000062 per degree Fahrenheit.

The specific gravity of cast iron varies from 6.9 to 7.5. It is usually taken at 7.22, corresponding to a weight of 450 lb. per cubic foot. In general, the specific gravity increases with the strength and the number of remeltings.

The shrinkage of cast iron in cooling varies according to its shape and purity, varying from $\frac{1}{32}$ in. per foot for large castings to $\frac{1}{8}$ in. per foot for bars. Pure cast iron will shrink about $\frac{3}{10}$ in. per foot of length.

When dropped on a concrete or stone floor, white cast iron has a bright metallic ring, malleable cast iron a dull ring, and gray cast iron a "dead" sound.

252. Allowable Working Stresses for Cast Iron.—Allowable unit working stresses for cast iron depend upon the character of the loads, grade of metal, and size of casting. Large castings are weaker than small ones due to shrinkage, nonuniformity of structure, and chance for defects.

For variable loads, the allowable working stresses in tension are 3,000 lb. per square inch; in compression, 16,000 lb. per square inch; and in shear about 2,500 lb. per square inch. Working stresses for steady loads may be taken about 25 per cent greater than those for variable loads. For repetitive loads, changing alternately from tension to compression, the allowable unit working stress should not be over 1,000 lb. per square inch. Cast iron is too brittle to be used to resist shocks or impact loads.

253. Uses of Cast Iron.—For structural purposes, cast iron is not used to so large an extent as wrought iron and steel. Cast iron can be used for posts, columns, column bases and caps, bearing plates, etc.

Cast iron is much used for parts of machines because it is cheap and can be cast in almost any form. In places where cast iron can be used, it suffers but little from competition by other metals.

F. MALLEABLE CAST IRON

254. Definition of Malleable Cast Iron.—Malleable cast iron is annealed white cast iron in which the carbon has been separated from the iron without forming flakes of graphite as in the gray cast iron.

255. Making the Castings for Malleable Cast Iron.—White pig iron, containing not more than 0.60 per cent of manganese, not more than 0.22 per cent of phosphorus, and not more than 0.05 per cent of sulphur, is used in the manufacture of malleable cast iron. The total carbon content must be more than 2.75 per cent. The amount of silicon varies inversely with the size of the casting; heavy castings require about 1.0 per cent, and light castings about 2.0 per cent of silicon. Sometimes malleable scrap (never over 20 per cent), steel scrap (never over 10 per cent), or wrought-iron scrap (not over 5 per cent) is mixed with the pig iron. Malleable scrap is hard to melt, but it increases the strength of the castings. Steel scrap and wrought-iron scrap tend to make the castings stronger.

The white pig iron, with the scrap, is melted in a cupola, an air furnace, or an open-hearth furnace; the air furnace being generally used. The molds are usually green sand molds. The molten white iron must be poured while it is very hot, and the pouring must be done very rapidly in order to secure good castings. After cooling, the castings are cleaned by any ordinary foundry method, and the imperfect ones rejected.

256. Annealing the Castings for Malleable Cast Iron.—The castings, which are to be annealed, are placed in annealing pots in such a way that they will not be deformed by any slight settlements. The annealing pots are made of cast iron, and are about 18 by 24 in. in cross section and from 15 to 48 in. high.

Iron scale, Fe_2O_3, is packed in the pots and around the castings. Sometimes the slag squeezed from wrought-iron puddle balls, powdered hematite ore, or magnetite is used instead of the iron scale. This iron scale acts as a decarburizer. Excess space in the annealing pots may be filled with good clean silica sand.

The annealing pots are placed in an oven and the temperature raised to a cherry red heat, about 1,450 degrees Fahrenheit, and held there from 3 to 5 days, depending on the size of the castings and the amount of decarburizing desired. Then the furnace is allowed to cool slowly for a few days before the castings are removed and cleaned.

To test the quality of the annealing, test plugs or small projections about ½ by ¾ in. by 1 in. long are cast on the more important pieces. These are broken off and the fracture examined. If properly annealed, the interior of the fracture should have a black velvety surface surrounded by a band of dark gray about 1/16 in. thick. This dark gray band should in turn be surrounded by a thin white band about 1/64 in. thick.

The effect of the annealing process is to change nearly all of the carbon from the combined form to a free amorphous form called "temper" carbon, to make the castings "malleable," and to about double their tensile strength. The decarburizer (packing of iron scale) prevents the oxidation and warping of the castings, and also extracts a large part of the carbon from the surfaces of the castings to a depth varying from 1/16 to ⅛ in.

257. Properties of Malleable Cast Iron. *Structure.*—The outermost skin, which is white in color and about 1/64 in. thick, consists of ferrite with a few impurities. The dark-gray layer, from ⅛ to 1/16 in. thick, consists of ferrite with a few scattered particles of temper carbon. This layer is much stronger and more ductile than the inner portion. The black interior consists of ferrite and many particles of temper carbon.

The tensile strength is the best criterion of the quality of malleable cast iron. Its tensile strength is about 45,000 lb. per square inch, with a yield point at about 70 per cent of the ultimate. The percentage of elongation varies from 2 to 7 per cent over a gage length of 2 in.

The compressive strength of small prisms of good malleable iron (load applied parallel to skin) will vary from 100,000 to 150,000 lb. per sq. in.

The average transverse strength (modulus of rupture) is about 60,000 lb. per square inch. Tests have given results varying from 54,000 to 90,000 lb. per square inch for a 1 in. square bar over a span of 12 in. The deflection varied from ½ to 1½ in.

Malleable cast iron is quite tough, and is much more able to withstand shocks and blows than the cast iron from which it is made.

Malleable castings can be bent when cold, and can be forged and welded to a greater or less extent.

258. Uses of Malleable Cast Iron.—Malleable cast iron has no structural uses, but it is used very much in the manufacture of articles which are too complicated in form to be readily forged, and which must be tougher and stronger than gray cast iron. Malleable cast iron has all of the advantages of gray cast iron, with respect to casting in various shapes, plus toughness, ductility, and strength nearly equal to that of some steels. Only cast or forged steel can compete with malleable cast iron in many of its uses, and then the malleable cast iron is usually cheaper.

Malleable cast iron is used in the manufacture of machinery parts, pipe fittings, small pinions, parts of railway rolling stock such as journal boxes, brake fittings, etc., and for various kinds of hardware, etc.

CHAPTER XII

WROUGHT IRON

A. DEFINITION AND CLASSIFICATIONS

259. Definition of Wrought Iron.—Wrought iron may be defined as nearly pure iron intermingled with more or less slag. The name "wrought iron" is applied to that commercial form of iron which is obtained by refining pig iron at a temperature not high enough to keep the metal in a molten state after the oxidation of the impurities, but just high enough to keep the iron in a pasty condition.

Wrought iron is made in a reverberatory furnace. It is composed principally of pure iron (ferrite) and slag (iron silicate) together with small amounts of impurities.

260. Classifications of Wrought Iron.—Wrought iron may be classified according to the method of manufacture, or according to its use.

According to Method of Manufacture

1. *Charcoal Iron.*—Made by using charcoal fuel and a charcoal hearth. The purest grades of wrought iron are made by this method.
2. *Puddled Iron.*—Made by the ordinary wet puddling process of manufacture.
3. *Busheled Scrap.*—Made as described in Article 266.

According to Use

1. *Staybolt Iron.*—Made from puddled or charcoal iron. While not the strongest, it is the toughest and most ductile wrought iron. Good for forging and welding.
2. *Engine-bolt Iron.*—Made from the same material as the staybolt iron. It is a little stronger but less tough and ductile.
3. *Refined Bar Iron.*—Made from a mixture of muck bars and iron scrap. Less strong, ductile, tough, and forgeable than the engine-bolt iron.
4. *Wrought Iron Plate.*—Class A. Made from puddled iron. A strong hard iron, but less ductile and tough than the best grades of wrought iron.

5. *Wrought Iron Plate.*—Class B. Made from a mixture of puddled iron and scrap. Not so good as the class A plate. Neither class A nor class B plate should be used for forging or welding.

B. MANUFACTURE OF WROUGHT IRON

261. The Materials for the Wet-puddling Process.—Practically all of the wrought iron manufactured at the present time is made from pig iron and various kinds of scrap by the puddling process. There are two kinds of puddling processes—wet puddling and dry puddling. The latter process is rarely used.

The pig iron commonly used in the manufacture of wrought iron is forge pig. This pig iron contains from 1.0 to 1.5 per cent of silicon, from 0.25 to 1.25 per cent of manganese, less than 1.00 per cent of phosphorus, and less than 0.10 per cent of sulphur. A large amount of silicon is desired so as to form enough slag to cover the molten iron. The amount of manganese is not important as it is practically all removed in the manufacture. The phosphorus and sulphur must be kept low as they are not completely removed.

The "fettling" is composed of the strong basic iron oxides that are used to line the furnace hearth. Some of these fettling materials are—basic slag from the puddling furnace or reheating furnace, hammer slag or scale from a rolling mill, and hematite ore. Enough fettling is used to cover the hearth to a depth of about 5 in.

The fuel used is a bituminous coal that burns with a long flame.

262. The Furnace Used in the Wet-puddling Process.—The furnace is of the reverberatory type and consists essentially of a puddling or working chamber (hearth), a grate or firebox at one end, and a stack at the other end. The furnace is made of masonry lined with firebrick. The outside is made of cast-iron plates fastened together with wrought-iron bolts. Iron castings support the hearth at some distance above the floor to allow for the circulation of air underneath. The hearth has a sloping roof which deflects the flame from the fire down upon the hearth. The working chamber is provided with openings for the purpose of charging and working the metal, and for the removal of the slag and the puddled iron. The fire grate has about one-third of the area of the hearth, and is separated from the hearth by a cast-iron air-cooled flue covered with a refractory material. Sometimes a steam jet is used to provide more air for the fuel and for the oxidation of the impurities. Dampers, etc. are provided for the regulation of the fire. The stack is connected with the hearth by suitable flue and draft openings, and is also provided with dampers to aid in the regulation of the fire.

263. Operation of the Furnace Used in the Wet-puddling Process.—The charge of the furnace consists of a large amount of fettling, which is uniformly compacted on the hearth, and about

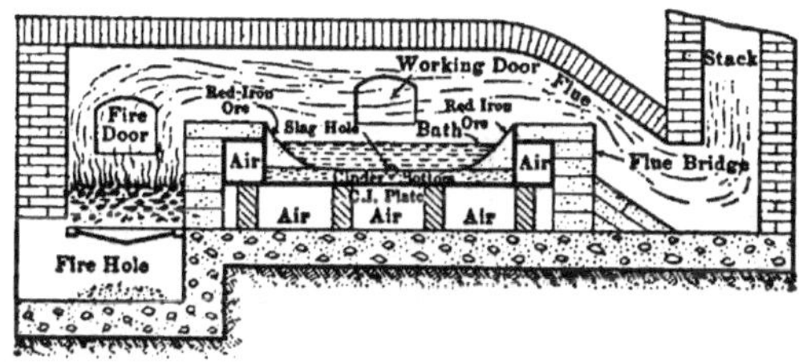

FIG. 89.—Section of a puddling furnace.

500 lb. of gray forge iron placed on top of the fettling. After a melting temperature has been reached, the reduction of the impurities occurs in four different stages:

1. The "melting down" stage lasts about half an hour after the fire is started, by the end of which time the iron will have melted and nearly all of the silicon and the manganese together with part of the phosphorus and a little of the sulphur will have been oxidized. These oxides leave the metal and join the slag.

2. The "clearing" stage lasts about 10 minutes, and in this stage the remainder of the silicon and of the manganese is oxidized together with a further quantity of phosphorus and sulphur. At this time it is usually necessary to add more red oxide of iron to make the slag more basic, and also to reduce the temperature of the furnace somewhat so that the carbon will not be oxidized before the phosphorus and the sulphur. Vigorous stirring or "rabbling" at this stage tends to help the oxidation.

3. During the "boiling" stage practically all of the carbon and most of the remaining phosphorus and sulphur are oxidized. The iron oxide in the slag unites with the carbon in the pig iron, producing carbonic oxide and iron. The iron combines with the iron on the hearth, while the carbonic oxide gas rises and causes

the molten metal to swell up and boil. When this gas comes through the surface of the bath, it burns in small flames of a light-blue color. The slag must be strongly basic in order to keep the phosphorus and sulphur in solution. During this stage the slag sometimes boils over the edge of the hearth and is caught and removed by means of a slag buggy. The puddler vigorously stirs or rabbles the charge to prevent the iron from oxidizing, to keep it from settling on the hearth, and also to secure a uniform product. Finally, the iron ceases to boil and "comes to nature" forming a spongy mass of metal which rests on the bed of slag.

4. During the "balling" stage, which occupies about 20 minutes, the temperature of the furnace is lowered and the

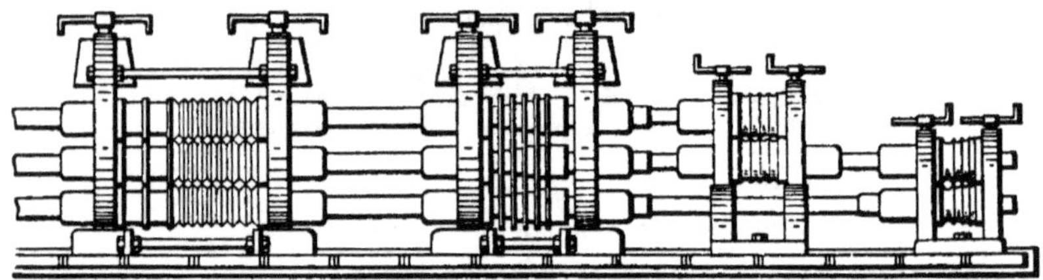

Fig. 90.—Showing principle of a merchant bar mill.

pasty mass of iron is divided into portions small enough to be removed from the furnace by the puddler. Before removing the pasty iron from the furnace, the puddler works each portion into a ball, storing the balls under the protection of the fire bridge to prevent the oxidation of the iron before the balls are taken from the furnace. Each ball weighs about 100 lb.

264. The Dry-puddling Process.—In the dry-puddling process, white pig iron is charged to the furnace and subjected to the action of an oxidizing flame. In this process, the necessary oxygen is supplied by the furnace instead of by the fettling.

265. Mechanical Treatment of the Puddle Balls.—*Squeezing or Shingling.*—When the puddle balls are removed from the furnace, they contain much slag. This slag is removed to a large extent by squeezing the balls in a squeezer, which reduces the diameter of the balls about one-half. Sometimes a ball is placed under a steam hammer and the slag pounded or "shingled" out. The squeezing or shingling causes a rise in the temperature of the ball which tends to melt the slag and thus aid in its removal.

Rolling Muck Bars.—After the squeezing, the balls are taken to a rolling mill and rolled and cut into rectangular bars called "muck" bars. The rolling usually removes a little of the remaining slag.

Reheating and Rerolling.—The muck bars are piled, tied in bundles with wire, reheated to a welding heat, and rerolled in commercial shapes (merchant bars). Sometimes the operations of piling, tying, reheating, and rerolling are repeated several times, resulting in an improvement in the quality and an increase in the strength of the bars. Not much advantage is gained by reheating and rerolling more than three times. The shapes of the commercial bars are sheets, plates, strips, bars, round and square rods, angle irons, tee irons, channels, I-beams, Z-bars, etc.

266. Wrought Iron Made from Scrap.—Sometimes wrought-iron scrap is tied together, heated, and rolled into commercial shapes. Another way is to make a box out of muck bars, fill it with scrap, heat it, and roll it. If the pieces of scrap are too small or too irregular to be tied together as muck bars are, they may be "busheled" together, placed in a small furnace, and treated as an ordinary puddle ball. All of these methods of making wrought iron from scrap result in an inferior product.

If some steel scrap is heated with the muck bars, the resulting product will have many of the properties of soft steel.

267. Defects in Wrought Iron.—The principal defects in wrought iron are rough edges, spilly places, blisters, and excess of slag. Rough edges are due to careless workmanship, imperfect rolls, or red shortness. Spilly places are spongy or irregularly spotted parts, and are generally caused by imperfect puddling. Blisters are caused by the presence of gas, probably carbonic oxide, in the iron when it is being rolled. An excess of slag is due to imperfect or insufficient squeezing, forging, or rolling of the balls and bars.

C. CONSTITUTION, PROPERTIES, AND USES OF WROUGHT IRON

268. Composition and Constitution of Wrought Iron.—The following table gives the chemical composition of some kinds of wrought iron. From this table it is seen that wrought iron is nearly all pure iron with a little slag. The strength of wrought iron is affected to some extent by its chemical composition, an increase in the carbon content causing an increase in strength.

CHEMICAL COMPOSITION OF WROUGHT IRON

Chemical element	Common wrought iron, per cent	Best wrought iron, per cent	Swedish wrought iron, per cent
Carbon	0.05 to 0.10	0.06	0.050
Phosphorus	0.18 to 0.35	0.15	0.055
Sulphur	0.04 to 0.06	0.03	0.007
Silicon	0.20 to 0.23	0.20	0.015
Manganese	about 0.10	0.06	0.006
Slag	2.80 to 3.10	2.80	0.610

Swedish wrought iron is a very pure wrought iron made in Sweden.

The constituents of wrought iron are ferrite (pure iron), slag (silicates and phosphates of iron and manganese), and a little pearlite due to the presence of the carbon. The little carbon in wrought iron combines with some of the iron to form cementite, which in turn combines with ferrite to form pearlite.

The size of the crystalline grains of ferrite in the wrought iron depends on the temperature from which the hot iron is cooled, the length of time held at that temperature, the rate of cooling, the mechanical working during the cooling, and the temperature at which this working is stopped. High temperatures and slow cooling both tend to increase the size of the crystals. Mechanical working tends to overcome the bad effects of coarse crystals by breaking up the large crystals and retarding their formation and growth.

269. Tensile Strength of Wrought Iron.—The tensile strength of wrought iron depends upon the direction of the load in regard to the grain or fibers of the wrought iron. The strength across the fibers is from 60 to 90 per cent of the strength along the fibers.

In regard to the effect of the amount of reduction in size due to the rolling, tests have shown that, if the ratio of the finished size of bar to the size of pile of muck bars is kept constant, practically the same tensile properties are shown by all sizes of wrought-iron rods.

The effect of previous straining or cold working on the tensile properties of wrought iron is to raise the elastic limit and ultimate strength and to decrease the elongation.

Annealing removes the effect of overstrain and also tends to lower the elastic limit and ultimate strength of the original bar.

The following values are average results from tensile tests on good grades of wrought iron:

Elastic limit	About 25,000 lb. per square inch
Yield point	About 30,000 lb. per square inch
Ultimate strength	About 50,000 lb. per square inch
Elongation in 8 in	About 20 per cent
Reduction in area	About 30 per cent
Modulus of elasticity	About 27,000,000 lb. per square inch

270. Compressive Strength of Wrought Iron.—The compressive strength of wrought iron depends upon the same factors as the tensile strength. The values obtained in compression tests are about the same as those obtained in tension tests. The ultimate strength may be taken as varying from 45,000 to 60,000 lb. per square inch, and the yield point from 25,000 to 35,000 lb. per square inch. Overstraining a wrought-iron bar in tension will impair its properties in compression, and *vice versa.*

271. Shearing Strength of Wrought Iron.—Wrought iron offers a greater resistance to shearing forces perpendicular to the fibers than it does to shearing forces parallel to the fibers. The following are general ranges of values for wrought iron in shear and torsion:

	Pounds Per Square Inch
Ultimate shearing strength parallel to the fibers	20,000 to 35,000
Ultimate shearing strength across the fibers	30,000 to 45,000
Elastic limit in torsion	17,000 to 25,000
Ultimate strength in torsion	45,000 to 60,000
Modulus of elasticity in torsion	about 12,500,000

272. Transverse Strength of Wrought Iron.—The transverse strength of wrought iron depends upon the same factors as do the tensile and compressive strengths. The results obtained in cross-bending tests are about the same as those obtained in tension tests. The yield point may be taken as varying from 25,000 to 35,000 lb. per square inch, and the ultimate strength from 40,000 to 60,000 lb. per square inch.

273. Fracture of Wrought Iron.—When good wrought iron is broken in tension or bending, the fracture is fibrous, and darker in color, more rough, and more jagged than that of mild steel. This is largely due to the presence of the slag in the iron. If the wrought iron is broken very suddenly, as by shocks, sudden blows, or impact, the fracture is more crystalline and granular in appearance.

274. Welding of Wrought Iron.—Welding is one of the most important properties that wrought iron possesses. It is the joining together of two pieces of the iron by pressing or hammering them while at a very high temperature, but which is not high enough to melt the iron. The ease with which wrought iron is

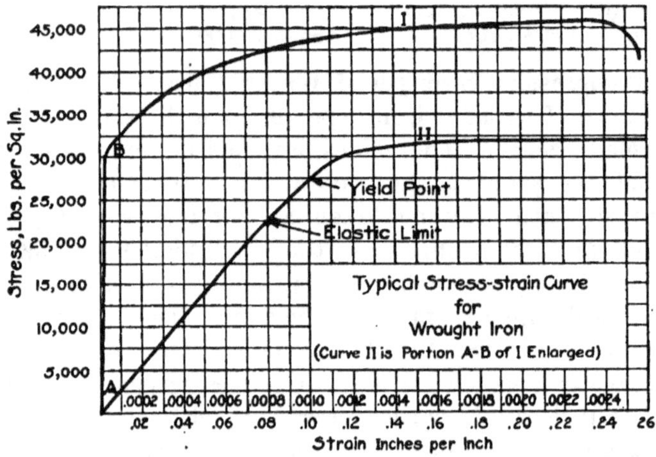

Fig. 91.—Typical stress-strain curve for high grade wrought iron in tension.

welded is due to two things—the absence of a large percentage of impurities such as carbon, silicon, and sulphur, and the ability of the wrought iron to remain in a plastic state (a white heat) through a considerable range of temperature.

It is very difficult to make a good welded joint because of the formation of iron oxide (melted slag) at the joint. In order to remove as much as possible of this melted slag in the welding, the surfaces of the two pieces of iron that are to be brought together should be convex. The use of a flux, such as borax, in welding aids the work by making the slag more soluble and, therefore, easier to remove from the joint.

Care should be taken not to leave the iron next to the joint in a brittle condition. This brittleness, which is caused by coarse crystals, may be remedied by working the wrought iron under a hammer or press until the critical range of temperature has been passed.

The welding temperature in practice is about 2,400 degrees Fahrenheit, while the critical temperature is about 1,275 degrees Fahrenheit.

The strength of welded joints varies from 30 to 90 per cent of the parts that have been joined.

275. Miscellaneous Properties of Wrought **Iron.**—Wrought iron has the important properties of toughness, ductility, malleability, and weldability, but it cannot be tempered.

The coefficient of expansion is about 0.0000065 per degree Fahrenheit.

The melting temperature is about 2,800 degrees Fahrenheit.

The specific heat is 0.0114.

The average specific gravity is 7.7.

Wrought iron containing much sulphur is "hot short" or "red short," that is, the iron is brittle and liable to break when worked at a red heat.

Wrought iron containing much phosphorus is "cold short." It has low ductility when cold, breaks with a crystalline fracture, and is unable to resist impact stresses.

276. Tensile Strength and Ductility Requirements for Wrought **Iron.**—The following requirements for tensile strength, elongation, and reduction in area for the different grades of wrought iron were taken from the American Society for Testing Materials Specifications for Wrought Iron.

WROUGHT IRON SPECIFICATION REQUIREMENTS

Kind of iron	Yield point, pounds per square inch	Ultimate strength, pounds per square inch	Elongation in 8 in., per cent	Reduction of area, per cent
Staybolt iron	29,400 to 31,800	49,000 to 53,000	30	48
Engine-bolt iron	30,000 to 32,400	50,000 to 54,000	25	40
Refined bar iron	25,000	48,000	22	
Wrought-iron plate				
6 to 24 in. wide, Grade A	26,000	49,000	16	
6 to 24 in. wide, Grade B	26,000	48,000	14	
24 to 90 in. wide, Grade A	26,000	48,000	12	
24 to 90 in. wide, Grade B	26,000	47,000	10	

277. Working Stresses for Wrought **Iron.**—It has been found that a stress in wrought iron in excess of the elastic limit causes a permanent set and raises the elastic limit. Repeated stresses above the elastic limit will cause a loss of strength and even failure if repeated a large number of times. Consequently, the working stresses for wrought iron should never exceed the elastic limit.

The following values of the allowable unit working stresses are for an average grade of wrought iron and should be increased about 30 per cent for the very best grades. This table was taken from the "American Civil Engineer's Pocket Book." All values are given in pounds per square inch.

UNIT WORKING STRESSES FOR AVERAGE WROUGHT IRON

Kind of load	Steady stress	Variable stress	Shocks and impact
Tension	14,000	10,000	4,000
Compression	13,000	9,000	3,000
Shear	10,000	7,000	3,500
Torsion	5,000	3,500	1,500
Cross bending	12,500	8,500	3,500

278. Uses of Wrought Iron.—Wrought iron is used for spikes, nails, bolts, nuts, wire, chains, chain rods, horseshoe bars, sheets, plates, staybolts, engine-bolts, pipes, tubing, third rails, armatures, electromagnets, and in the manufacture of crucible steel.

Before 1890, wrought iron was used very much in bridge and structural-building work, but since that date structural steel has replaced it. Structural steel costs less than wrought iron and is about 20 per cent stronger.

CHAPTER XIII

STEEL

A. DEFINITIONS AND CLASSIFICATIONS

279. Definitions of Steel.—The Committee of the International Association for Testing Materials has proposed the following definition: Steel is iron which is aggregated from pasty particles without subsequent fusing; is malleable at least on some one range of temperature; and contains enough carbon (more than 0.30 per cent) to harden usefully on rapid cooling from above its critical temperature.

The following definition proposed by Prof. H. M. Howe is, perhaps, better than the one given above. Steel is iron which is usefully malleable at least in some one range of temperature; and, in addition, either (*a*) is cast into an initially malleable mass; or (*b*) is capable of hardening greatly on sudden cooling; or (*c*) is both so cast and so capable of hardening.

280. Classifications of Steel.—Steel may be classified according to method of manufacture, use, or carbon content.

According to the method of manufacture, steel may be divided into cementation steel, crucible steel, acid Bessemer steel, basic Bessemer steel, acid open-hearth steel, basic open-hearth steel, duplex steel, electric steel, etc.

According to its use, steel may be divided into the following classes; structural-rivet steel, structural steel, boiler-rivet steel, boiler-plate steel, machinery steel, rail steel, gun steel, axle steel, spring steel, tool steel, cable-wire steel, etc.

Steel may also be classified into carbon or alloy (special) steels, depending on the chemical element which tends to control its strength. Carbon steels may be divided into soft, medium, hard, and very hard steel, according to the percentage of carbon present. Alloy or special steels include all other steels, except the carbon steels, and may be divided into many classes. The name of each class depends upon the name or names of the element or elements (other than iron or carbon) governing its distinctive properties. Some examples of alloy steels are: nickel steel, manganese steel, nickel-vanadium steel, vanadium steel, chrome steel, etc.

For general classifications of iron and steel see Art. 233 and Art. 234.

B. METHODS OF MANUFACTURE OF STEEL

281. The Cementation Process.—The principle of this process is the absorption of the carbon by wrought iron at a high red heat and thus converting the wrought iron into steel.

Alternate layers of charcoal and wrought-iron bars are packed in a "converting pot" (made of a refractory stone or brick) so that the charcoal completely surrounds each bar. Several of these pots are placed in a furnace and the temperature gradually raised to about 1,250 degrees Fahrenheit at the end of 3 days' time. The furnace is kept at this temperature from 7 to 12 days, depending on the carbon content desired in the steel. When the proper amount of carbon has been absorbed, the fires are drawn and the furnace allowed to cool slowly for about a week before the bars are removed.

The presence of a little slag in the wrought-iron bars causes the formation of carbon monoxide gas which makes "blisters" on the surface of the bars, hence the name "blister steel" has been given to this product.

On account of the cost and the length of time required, very little of this kind of steel is made at the present time.

282. The Crucible Process.—The principle of this process is the absorption, by molten wrought iron, of the carbon in such quantities as to change the wrought iron into steel.

About 80 lb. of wrought iron with a little charcoal and manganese is placed in a closed crucible made of some refractory material. Several of these crucibles are placed in a furnace and subjected to an intense heat which melts the metal in 2 or 3 hours. The crucibles are then kept in the furnace for about half an hour longer until the metal is "killed" (has ceased to boil and evolve gases). After the "killing," the crucibles are removed from the furnace and the molten metal poured into ingot molds. The time required for a heat varies from 3 to 5 hours.

As the cost of this kind of steel is quite high, it is used only in the manufacture of tools, cutlery, springs, projectiles, etc., where a steel of a high grade is required.

283. The Principle of the Bessemer Process and the Plant Equipment.—The principle of the Bessemer process is the oxidation of the carbon and some of the other impurities by blowing a blast of cold air through a bath of molten pig iron in a converter.

The essential parts of the plant are the blast furnaces or cupolas, mixers, converters, blowers and blowing engines, together with the necessary ladles, ingot molds, etc.

The molten pig iron is brought from the blast furnaces or cupolas and stored in a mixer until it is time to charge the

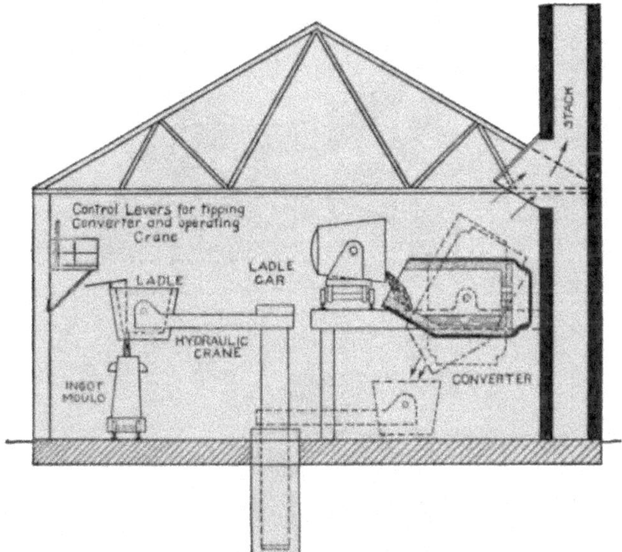

Fig. 92.—Diagram of a Bessemer steel plant.

converter. This mixer is a large steel reservoir lined with firebrick. It has a capacity of from 100 to 600 tons of molten pig iron and is used to keep the molten metal hot and also to allow the mixing of the iron from several blast furnaces so as to secure the desired grade of iron.

The converter is a bucket-shaped vessel of steel with an eccentric conical snout. Its capacity varies from 15 to 30 tons of metal. The converter is lined with a refractory material and is supported on trunnions so that it can be tipped. The bottom is pierced with a large number of holes through which the air blast enters. On account of the fact that the lining in the bottom burns out faster than that of the sides, the bottom is made so that it is detachable and can be removed and have its lining renewed

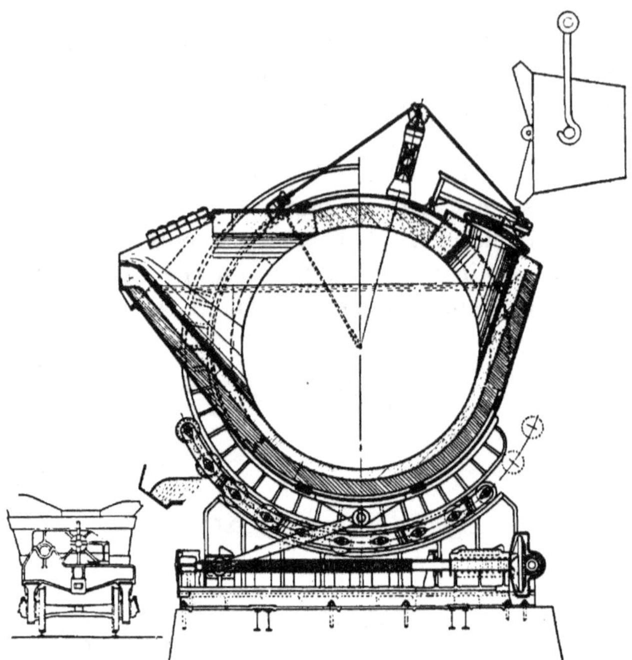

FIG. 93.—Cross section of a mixer.

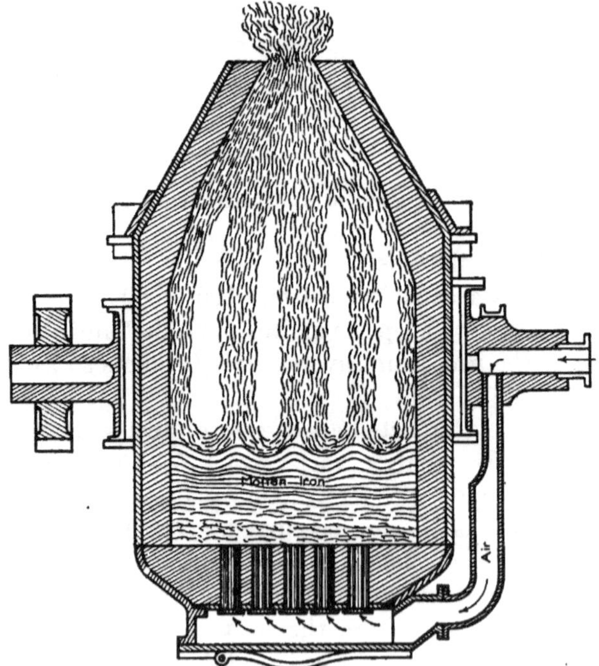

FIG. 94.—Section through Bessemer converter while blowing. (*Stoughton*).

whenever necessary. The average life of the bottom is only about 25 heats. The lining of the converter is of a siliceous brick for the acid process, and of a basic material such as dolomite or limestone for the basic process.

284. The Acid Bessemer Process.—About 15 or 20 tons of molten pig iron are brought in a ladle from the mixer and poured into the converter which is tilted to a horizontal position to receive the charge. Then the converter is rotated to a vertical position and the air blast turned on. This air blast, under a

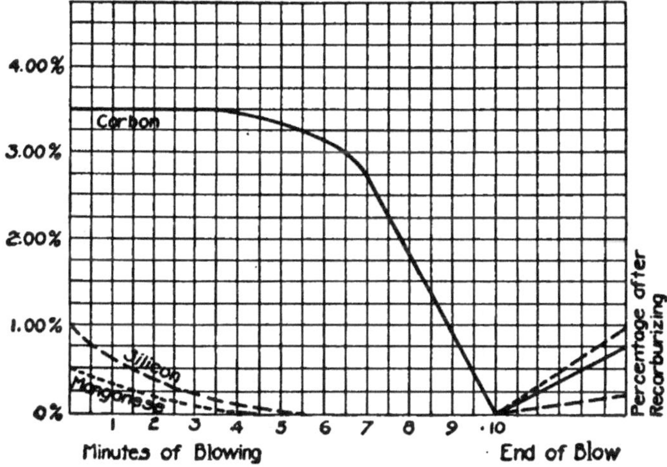

Fig. 95.—Removal of impurities in acid Bessemer process. (*Bradley Stoughton.*)

pressure of about 25 lb. per square inch, is forced through the molten iron for a period of about 10 minutes, after which time the impurities will have been nearly all removed by oxidation.

The silicon is oxidized first with the evolution of much heat, and is followed by the oxidation of the manganese. These reactions produce an acid slag which floats on the top of the metal. The temperature of the molten iron now becomes so high that the carbon is oxidized, producing a brilliant flame that extends 20 or 30 ft. above the mouth of the converter.

When the carbon is all oxidized, the flame drops, the blast is turned off, and the converter is rotated to a horizontal position to receive a predetermined amount of speigeleisen or ferromanganese which causes the metal to boil. At this time the carbon passes into the metal while the manganese takes the oxygen from the metal and passes into the slag. The molten steel is then poured into a ladle from which the ingot molds are filled.

The length of time required for one heat is about 15 minutes.

In the acid Bessemer process the pig iron used must be low in phosphorus and sulphur because neither is eliminated, while the silicon content should be high in order to produce the necessary heat. The usual limits of the composition of the pig iron are: 1.0 to 2.0 per cent of silicon, 0.4 to 0.8 per cent of manganese, 3.5 to 4.0 per cent of carbon, 0.07 to 0.10 per cent of phosphorus, and 0.02 to 0.07 per cent of sulphur.

285. The Basic Bessemer Process.—In this process a little limestone flux is added to the charge to produce a basic slag that will dissolve the silica and thus prevent it from attacking the

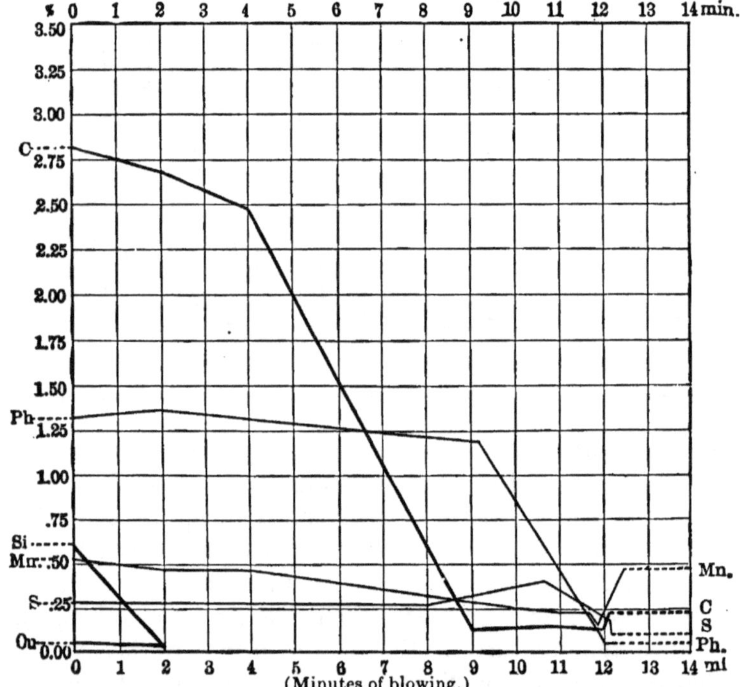

FIG. 96.—Removal of impurities in basic Bessemer process. (*Bradley Stoughton.*)

converter lining. The impurities are oxidized in the same order as in the acid Bessemer process until the carbon flame drops, after which the phosphorus oxidizes and dissolves in the slag with the evolution of a large amount of heat. To accomplish this, the blast must be turned on for about 6 minutes longer than in the acid process. If some manganese is present, the sulphur will also be eliminated from the iron. In order to keep the phosphorus from returning into the metal, the recarburizer (speigeleisen or ferromanganese) should not be added until the metal has been poured in a ladle and the slag skimmed off. The total length of time required for a heat is about 25 or 30 minutes.

In the basic process the sulphur and phosphorus can be eliminated and, consequently, a pig iron high in sulphur and phosphorus can be used. In fact, a high-phosphorus content is

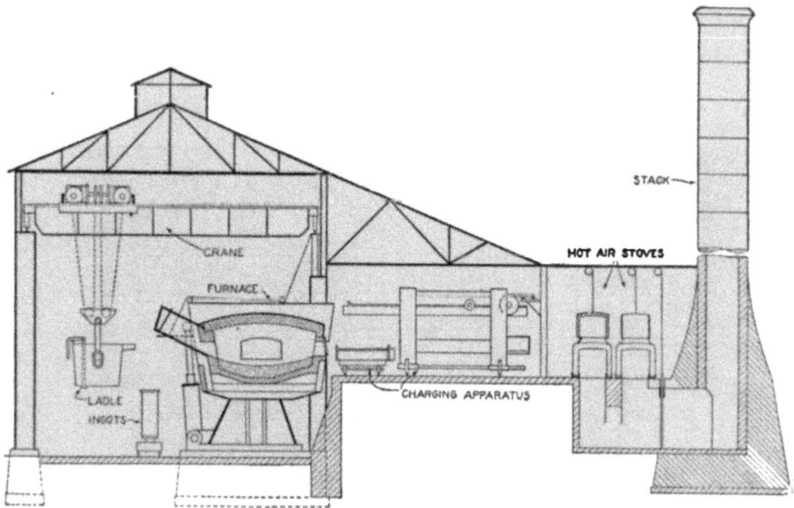

Fig. 97.—Diagram of an open-hearth steel plant.

desirable in order to produce the necessary heat. The limits of the composition of the average pig iron used are: 1.0 to 3.0 per cent of phosphorus, 0.2 to 1.0 per cent of silicon, 0.02 to 0.30 per cent of sulphur, 0.3 to 2.0 per cent of manganese, and 2.75 to 3.5 per cent of carbon.

286. The Principle of the Open-hearth Process and the Plant Equipment.—The principle of this process is the oxidizing of the impurities in the metal under the direct action of an oxidizing flame of gas and air burned in a reverberatory regenerative furnace.

The most essential part is the furnace. Some of the other essential parts are: ladles, overhead cranes, charging machines, gas producers, regenerators (usually included with the furnace), ingot molds, etc.

The essential parts of a reverberatory regenerative furnace are:

1. A large shallow hearth lined with a refractory material. This material must be basic or acid in character, depending on the kind of slag formed.

2. A cover over the hearth so shaped that the heat produced by the combustion of the gas and air will be deflected upon the hearth.

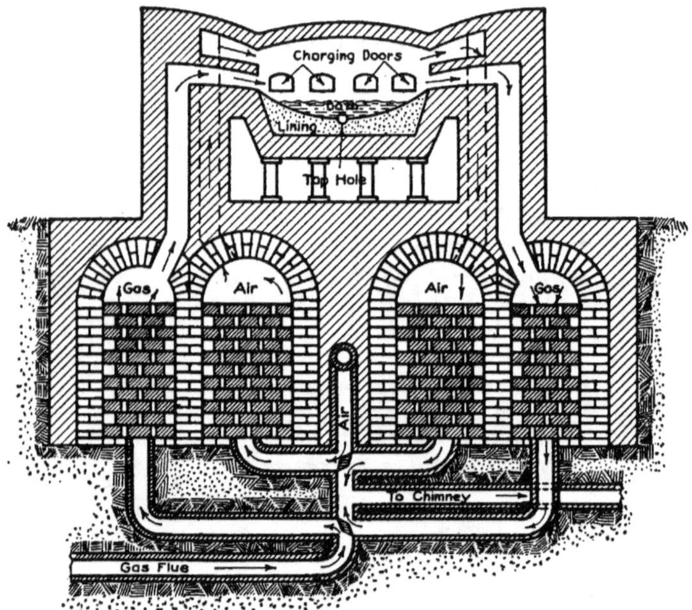

FIG. 98.—Section of an open-hearth furnace showing principle of operation.

3. Four or more systems of brick checkerwork (regenerators); two for heating the gas and air while the other two are being heated by the escaping burned gases.

4. A system of valves and passageways so that the admission of the gas and air may be controlled.

5. A charging door for adding the charge to the furnace.

6. A tap hole or spout for drawing off the molten metal.

The furnace is called "reverberatory" because the flame is deflected upon the hearth by the cover, and it is called "regenerative" because the escaping burned gases are used to heat the brick checkerwork.

Two kinds of open-hearth furnaces are used: (*a*) the stationary type from which the molten metal is drawn from a tap hole; and (*b*) the rolling furnace which can be rolled or tilted so that the molten metal may be discharged through a spout.

For fuel, a producer gas is used. This gas consists principally of nitrogen, carbonic oxide, and hydrogen. It is made by forcing air through a bed of red hot bituminous coal. Often a natural gas is used for fuel.

287. The Acid Open-hearth **Process.**—The charge, consisting of about one-third pig iron and two-thirds steel scrap, is placed on the hearth. The gas and air are turned on and the temperature gradually raised until the charge is melting at the desired rate. At the end of 5 or 6 hours, the silicon, manganese, and a large part of the carbon will have been oxidized out of the metal.

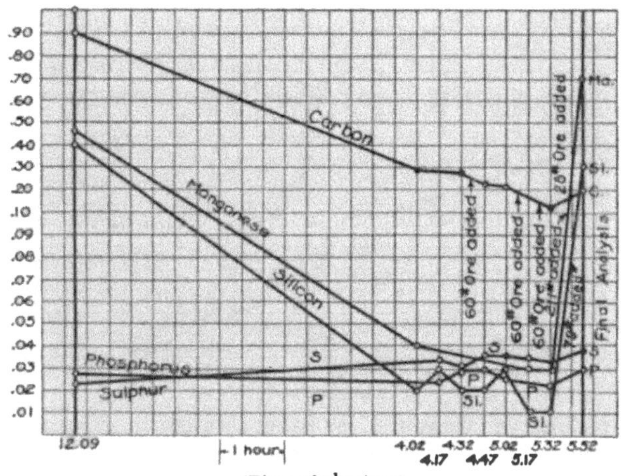

FIG. 99.—Removal of impurities in acid open-hearth process. (*Bradley Stoughton.*)

Then a small amount of the metal is drawn from the furnace and a small test bar made, from whose fracture the carbon content is estimated. If the carbon content is high, some ore is added to produce oxidation; if the carbon content is low, some more pig iron is added. Just before the metal is drawn off, a little ferromanganese is added to remove the oxygen from the metal and to stop any further oxidation of the carbon. The charge is then conveyed from the furnace and poured into ingot molds. The time required for running a heat varies from 6 to 10 hours.

The composition of the pig iron used is: 0.8 to 2.0 per cent of silicon, 0.3 to 0.5 per cent of manganese, less than 0.05 per cent of phosphorus, less than 0.05 per cent of sulphur, and from 3.0 to 4.0 per cent of carbon. The composition of the steel scrap used is: 0.1 to 0.3 per cent of silicon, 0.4 to 0.8 per cent of manganese, less than 0.05 per cent of sulphur, less than 0.05 per cent of phosphorus, and from 0.2 to 0.3 per cent of carbon.

288. The Basic Open-hearth Process.—The charge in this basic process consists of about equal parts of pig iron and steel scrap, together with a little ore and limestone flux. About 3 or 4 hours are required to melt the charge and oxidize most of the silicon and some of the manganese and carbon. Then

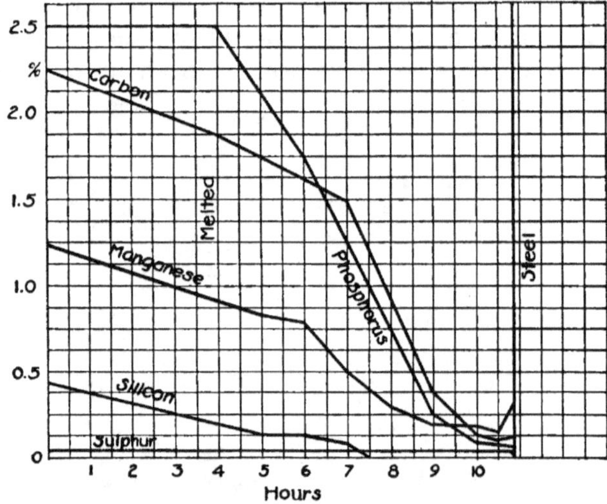

FIG. 100.—Removal of impurities in basic open-hearth process. (*Bradley Stoughton.*)

the phosphorus is oxidized and absorbed by the slag, while some of the manganese combines with some of the sulphur and is dissolved in the slag. The charge is tested from time to time by molding and fracturing a small test bar. When the elements are reduced to the proper percentages, the molten steel is drawn off in a ladle. After the slag is skimmed off and a recarburizer added, as in the basic bessemer process, the molten steel is poured into the ingot molds. The total time required to run a heat varies from 8 to 12 hours.

The steel scrap used in this process has about the same composition as the steel scrap used in the acid open-hearth process, except that the phosphorus and sulphur contents may be a little higher. The composition of the pig iron used is: less than 1.0 per cent of silicon, more than 1.0 per cent of manganese, from 1.0 to 2.5 per cent of phosphorus, from 0.02 to 0.30 per cent of sulphur, and from 2.5 to 3.5 per cent of carbon.

Open-hearth steel, like Bessemer steel, is used mostly for all kinds of structural purposes.

289. The Electric Process.—The principle of this process is the same as that of the open-hearth process except that electricity, instead of gas and air, is used to produce the heat necessary for the oxidation of the impurities. In the electric process, no oxygen is required to supply the heat. This is an advantage over all other processes. The electric furnace is very efficient

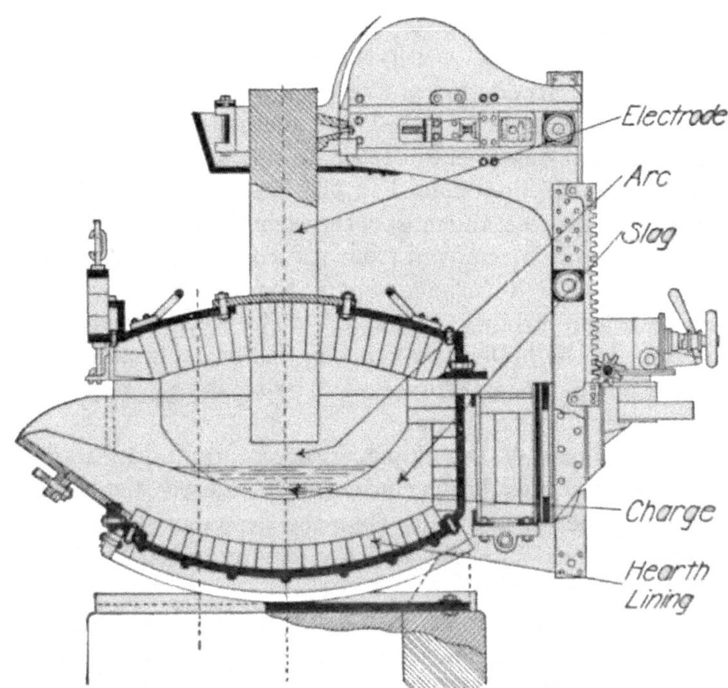

Fig. 101.—Section of a Heroult arc type electric furnace.

in removing the sulphur and the oxygen from the steel, but it is less efficient in removing the phosphorus.

There are three types of electric furnace:

1. Furnaces using an open arc between electrodes above the bath. The Stassano furnace is an example of this type.

2. Furnaces using an arc between the electrodes and the bath. The Giroud, Heroult, and Keller furnaces are examples of this kind.

3. Furnaces of the induction type in which the bath forms the secondary coil (or part of the coil) of a transformer. The Roehling-Rodenhauser and Kjellin furnaces are examples of this type.

The power consumption of an electric furnace varies from

150 to 1,000 kilowatt-hours per ton of steel produced according to the type of furnace, kind of materials in the charge, and the temerature of the charge at the start. The time necessary for a heat varies from 2 to 5 hours, depending upon conditions.

In the production of high quality steel, the electric furnace is a strong competitor of the crucible process, because it can produce larger quantities at a heat and at a slightly lower cost. The electric furnace has an advantage over all other furnaces in making special alloy steels, because it does not need to be operated under oxidizing conditions. However, the electric furnace cannot compete in cost with the bessemer and open-hearth processes in the production of steel of a medium or low quality.

290. The Duplex Process.—This process is usually a combination of the acid Bessemer and basic open-hearth processes. The molten pig iron is first placed in the Bessemer converter until the silicon, manganese, and most of the carbon are oxidized. Then the molten metal is removed from the converter and placed in a basic open-hearth furnace where the phosphorus and the remainder of the carbon are removed. The recarburizer is added to the steel in the ladle after the slag has been skimmed off, as in the basic open-hearth process. The total time required for a heat is about 6 or 8 hours.

The advantages of the duplex process are: a low-grade pig iron with a high phosphorus content may be used; a steel is produced that is better in quality than the Bessemer steel; the time required for a heat is about half the time required by the basic open-hearth process.

Another kind of duplex process is where the preliminary refining of the steel is done in a Bessemer converter or an open-hearth furnace and the final refining in an electric furnace. This process gives a steel of a high quality and at a lower cost than when the electric process is used alone.

291. The Triplex Process.—In this process the molten metal is taken from a mixer and placed in an acid Bessemer converter where the silicon, carbon, and manganese are nearly all removed. Then the molten metal is transferred to a basic open-hearth furnace where the phosphorus is removed and the steel recarburized. After this, the molten steel is placed in an electric furnace for the final refining; that is, for the removal of the sulphur and the oxygen. The triplex process gives a very high-grade steel and is less expensive than the electric process.

292. Comparison of the Different Processes.—

Quality	Cost	Length of time for one heat	Total quantity produced
1. Electric...........	Crucible	Basic open-hearth	Basic open-hearth
2. Triplex..................	Electric	Acid open-hearth	Acid Bessemer
3. Crucible.................	Triplex	Triplex	Basic Bessemer
4. Basic open-hearth.........	Acid open-hearth	Duplex (usual)	Acid open-hearth
5. Acid open-hearth..........	Basic open-hearth	Crucible	Crucible
6. Duplex (usual)	Duplex (usual)	Electric	Duplex (usual)
7. Basic Bessemer............	Basic Bessemer	Basic Bessemer	Electric
8. Acid Bessemer.............	Acid Bessemer	Acid Bessemer	Triplex

C. COMPLETING THE MANUFACTURE OF THE STEEL

293. Casting the Ingots.—Most of the steel is cast into ingots, which are about 7 ft. high, 18 in. square at the bottom, and about 15 in. square at the top.

The ingot molds are placed on small flat cars that run on a track. At the proper time the molten steel is drawn from the furnace into a ladle which is a bucket shaped vessel of steel lined with a refractory material and having a valve at the bottom. The cars carrying the ingot molds are run under the ladle, and each mold is filled by means of the valve. The cars then pass under cranes that remove the molds from the ingots. The molds are washed with clay water to prevent the steel from sticking to them, and are used again. The ingots pass on to a "soaking pit" or a reheating furnace. The average life of an ingot mold is about 100 casts.

294. Defects in Ingots.—When the ingots cool, they form a structure that is not very homogeneous as the metal on the outside cools first so that when the metal of the inside cools and shrinks, a cavity or *pipe* is formed inside the ingot and near its upper end. Oxides of phosphorus, sulphides of iron, and sulphides of manganese, together with some carbon, tend to *segregate* near the upper portion of the central part of the ingot. The gases and slag in the molten metal rise to the top of the ingot and form *blowholes*. *Ingotism* is the formation of large crystals of steel and is caused by slow cooling or by casting at too high a temperature.

Piping and segregation may be overcome by a proper proportioning of the different elements, while the only remedy for blowholes is to cut off the top of the ingot. Careful rolling and forging will remove the bad effects of ingotism by reducing the size of the steel crystals.

295. Reheating the Ingots.—For good working, it is necessary for the ingot to have the proper working temperature throughout. As the outside of the ingot cools more rapidly than the interior, it is necessary to place the ingot in a furnace where the outside of the ingot may be heated and kept at the proper temperature for working until the interior has solidified and cooled to that temperature. If the ingot becomes too cool during the working, it is necessary to stop and reheat it before going on with the working.

The furnaces used for reheating purposes are of three types: the "soaking pit," the regenerative gas-fired pit furnace, and the billet heating furnace.

The "soaking pit" is a pit in which the ingot is charged through the top and kept in a vertical position. The purpose of the soaking pit is to equalize the temperature of the ingot. No fuel is used, the interior of the hot ingots supplying the heat. The soaking pit is used to equalize the temperature of the ingot before the working is commenced.

The regenerative, gas-fired, pit furnace is a vertical furnace using gas for fuel. The ingot is charged through the top, and remains in the furnace until it is heated to the proper working temperature. This furnace can be used for equalizing the temperature of the ingot and also for reheating the ingot when it cools during the working.

The billet heating furnace is a horizontal gas-fired furnace in which the billets enter at the cool end of the furnace and are pushed through by means of a hydraulic ram, discharging from the hot end. This furnace is used for heating small pieces of steel either before or during the working.

296. Rolling.—The rolling is done by rolling mills which consist essentially of two (two high) or three (three high) layers of smooth chilled cast-iron rollers of the desired form. Sometimes a mill has a set of vertical rollers, placed just outside the horizontal ones, which keep the edges of the metal smooth but do not reduce it. Such a mill is called a universal mill.

The ingot passes through the rolls many times (from 2 to 20) before emerging in the desired form, each passage through the rolls changing the shape of the steel somewhat. The action of the rollers compresses the metal under the rollers and also tends to produce longitudinal tension in the surface fibers of the steel. The speed of rolling is quite great, varying from a few miles per hour for special shapes, to about 10 miles per hour for rails, and to as high as 30 miles per hour for rods. Too great a speed in rolling causes a heating of the steel, due to the rapid distortion. Rolling

Fig. 102.—"Three High" I-beam roughing rolls.

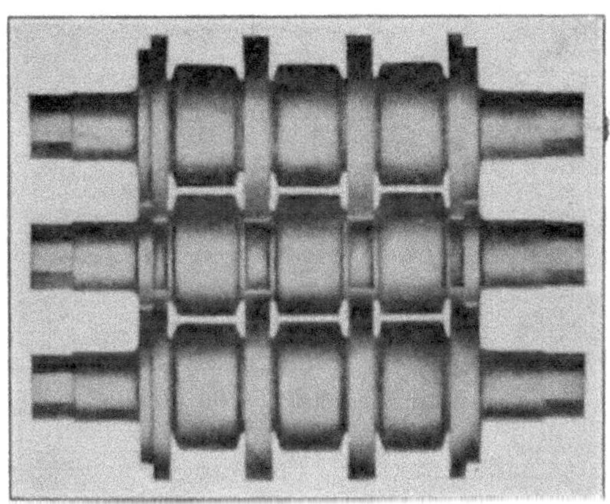

Fig. 103.—"Three High" I-beam finishing rolls.

mills produce plates, bars, structural steel shapes, rails, and rods.

297. Forging and Pressing.—Forging steel is working it under a hammer until it is of the desired size and shape. Most of the hammers used at the present time are steam hammers, and they vary in size from a few hundred pounds up to 30 or 50 tons.

Drop forgings are forgings made by using dies in connection with the steam hammer. The metal usually passes through a series of dies before becoming of the proper size, shape, and finish.

Forging may also be done by using a hydraulic press instead of a steam hammer. With the hydraulic press the force is applied more slowly and acts for an appreciable length of time, while with a hammer the force is applied as a blow. Hydraulic presses vary in size from a few tons pressure up to 1,400 tons pressure or more.

The mechanical treatment of an ingot, such as forging, pressing, drawing, or rolling, greatly improves the quality of the steel by solidifying the metal, reducing the blowholes, and increasing the

Fig. 104.—Rolling mills. (*Illinois Steel Co.*)

strength and specific gravity. For the best results, the work should be done when the ingot has cooled to a low-red heat. Forging works the metal better, but it is more expensive and less rapid in operation than rolling. Much more steel is made by rolling than by forging.

298. Wire Drawing.—In making wire, the steel is first rolled in rods and then these rods are drawn through holes in a plate. These holes vary in size from the diameter of the rod down to the diameter of the desired wire. Some thick lubricant is used to reduce friction and to prevent wear of the holes in the draw plate. The diameter of the rolled rods is usually from

¼ to ½ in. The sectional area of the rod is reduced about 20 or 25 per cent for each hole that it is pulled through. Cold drawing makes the metal very hard, and it must be annealed to a low-red heat after it has been drawn from 3 to 10 times. The number of drawings through the holes depends upon the original diameter of the rod and the desired diameter of the wire, and may be as many as 20 or more.

D. HEAT TREATMENT OF STEEL

299. Hardening of Steel.—The heat treatment of steel by hardening, tempering, and annealing greatly influences its physical properties. Every steel has a certain "critical" temperature in the range of which important molecular changes occur in heating and cooling. In general, this range is from a low-yellow heat down to a dull-red heat.

Steel is hardened by heating it up to this critical temperature, which is usually between 1,250 and 1,600 degrees Fahrenheit, and then retarding the molecular changes that occur in slow cooling by suddenly plunging the steel into molten lead, oil, water, ice water, or iced brine, etc., according to the degree of hardness desired. The hardness increases with the rate of cooling, and, in general, the cooler the quenching liquid, the harder the steel. The degree of hardening, as well as the critical temperature, also depends to some extent upon the amount of carbon, manganese, chromium, tungsten, and other elements present.

300. Tempering of Steel.—As some hardened steels are too brittle to use, they must be tempered to reduce the hardness to some extent. "Tempering" is accomplished by heating the steel up to a temperature which is less than the critical temperature, and then quenching it in some liquid as oil, water, etc. For most steels, the temperature for tempering varies from 425 to 600 degrees Fahrenheit. This temperature is indicated by the color of the film of oxide that forms on the surface of the steel. This color varies from a pale-yellow (about 425 degrees Fahrenheit) through straw, brown, purple, and blue to dark-blue (about 600 degrees Fahrenheit). Tempering the steel tends to increase its ultimate strength.

301. Annealing of Steel.—Annealing consists of heating the steel up to a light red heat and then allowing it to cool very slowly for some (2 or 3) days. The usual annealing temperatures are between 400 and 900 degrees Fahrenheit. If the size of grain is to be reduced much, the temperature must be raised slightly above the lower limit of the critical range of temperature. The heating must be done in such a way that the steel will not come in contact with the fuel and flames, and the pieces of steel must be supported so that they will not warp. Small pieces are often packed in charcoal in closed iron boxes.

Annealing removes any overstrain caused by the cooling or by the working of the steel; reduces the size of the grains of the steel; makes the steel soft and ductile; and reduces the elastic limit and ultimate strength.

302. Case Hardening of Steel.—Case hardening is accomplished by heating soft or medium steel in contact with carbon so that the carbon will penetrate the outer skin of the steel. The temperature required is about 1,650 degrees Fahrenheit. The time varies from 2 to 12 hours according to conditions. The penetration of the carbon is usually less than ¼ in.

Case hardening produces a high-carbon steel surface which is hard and which will resist wear, abrasion, cutting, and indentation, while the interior is left soft and tough and capable of resisting impact.

Harveyized armor plate is made by case hardening the side of an armor plate of tough medium steel.

E. STRUCTURE AND CONSTITUTION OF STEEL

303. Normal Constituents and Compounds.—The normal constituents of iron and steel are ferrite (pure iron), cementite (Fe_3C), and graphite (free amorphous carbon). These constituents appear in various forms and combinations when the iron or steel is in the solid and liquid forms, depending upon the amount of carbon, rate of cooling, presence of other elements, etc.

Molten steel is a solution of liquid carbide of iron in liquid iron whose carbon content is less than about 2.0 per cent (some authorities claim 1.7 per cent and others 2.2 per cent). If the carbon content were over 2.0 per cent, the liquid solution would be called molten cast iron. When the carbon is less than 2.0 per cent, there is no separation of the constituents when the solution cools, *i.e.*, it forms a "solid" solution. Steels are solid solutions.

When a molten mixture cools, one constituent may solidify

before another, or before the solution freezes as a whole. If the proportioning of the parts is such that the freezing point of the solution is reached before that of any of its parts, it is called a "eutectic" solution, in the case of the cast irons, and a "eutectoid" solution in the case of the steels. The substance formed

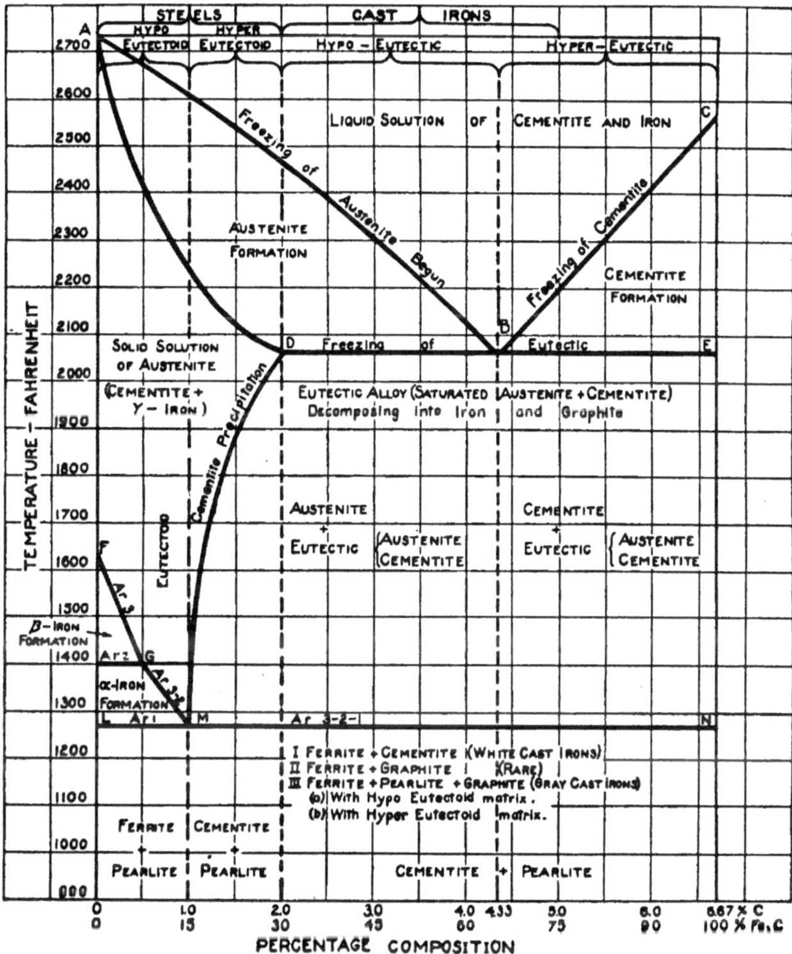

FIG. 105.—Roberts-Austen iron-carbon diagram. (Slightly modified.)

by the freezing of the eutectic solution is called the "eutectic," while that formed by the freezing of the eutectoid solution is called the "eutectoid." If there is an excess of iron carbide (Fe_3C) present, the steel solution is called "hyper-eutectoid" and the cast-iron solution "hyper-eutectic." But if there is an excess of free iron present, the steel or solid solution is called "hypo-eutectoid" and the cast-iron solution "hypo-eutectic."

304. Critical Temperatures.—It has been found that when a bar of soft steel containing about 0.20 per cent of carbon is gradually heated, the temperature of the bar increases regularly with the time up to a certain point, Ac_1, at which the increase in temperature is retarded for a short time. After this the temperature will again increase regularly until a second point, Ac_2, is reached, where a similar retardation will occur. If the heating is continued, a third such point, Ac_3, will be found before the steel melts.

In cooling, the same thing will happen but in the reverse order, the "critical" points found being called Ar_3, Ar_2, and Ar_1. These latter points are each about 55 degrees Fahrenheit lower than the corresponding points obtained when the bar is heated.

If the carbon content is increased, the point Ar_3 tends to approach point Ar_2 and will finally coincide with it. A further increase in the carbon content will tend to make the point Ar_{2-3} approach the point Ar_1, and, if the carbon is increased sufficiently, all three of these points will coincide. This is also true in regard to the points Ac_3, Ac_2, and Ac_1.

When a high-carbon steel is gradually cooled, the temperature will decrease regularly until the point Ar_{1-2-3} is reached, and then there will be a sudden momentary increase in temperature, after which the regular rate of cooling will be resumed. This phenomenon is called "recalescence," and the point the "recalescence point."

An explanation of this phenomenon is that there is a change in the molecular structure of the steel at each of these critical points. The iron above the Ar_3 point is called gamma iron; the iron between the Ar_3 and the Ar_2 points, beta iron; and the iron between the Ar_2 and the Ar_1 points, alpha iron.

The Ar_1 point is at about 1,275 degrees Fahrenheit, while the Ar_2 point varies between 1,400 to 1,275 degrees Fahrenheit, and the Ar_3 point from 1,650 to 1,400 degrees Fahrenheit (after which it coincides with Ar_2) according to the carbon content. Increasing the carbon content lowers Ar_3 until it coincides with Ar_2, and increasing the carbon content still further causes Ar_{2-3} to be lowered until it coincides with Ar_1.

Alpha iron is soft, ductile, and magnetic, while beta iron is hard, glassy, brittle, and non-magnetic.

305. Slow Cooling of Molten Steel.—When steel cools slowly, it passes from a liquid solution of iron carbide in gamma iron through austenite, martensite, troostite, and sorbite to pearlite. It will be all pearlite if the solution is a eutectoid solution.

If there is an excess of gamma iron present (hypo-eutectoid), this excess will gradually separate out in the form of ferrite before the pearlite is formed. This ferrite passes from gamma iron to beta iron and then into alpha iron. At the Ar_1 point, pearlite is formed, and the excess iron will be in the form of ferrite (probably a mixture of alpha and beta irons). Thus the resulting steel will be a mixture of ferrite and pearlite.

If there is an excess of iron carbide in the solution, this excess will separate out as cementite until the solution is eutectoid in character and until the point Ar_1 is reached, when the remainder of the solution will freeze as pearlite, the resulting steel being a mixture of pearlite and cementite.

Austenite is a solid solution of iron carbide (cementite) and gamma iron. It is stable above the Ar_1 point, may contain carbon up to about 2.0 per cent, and is unmagnetic, ductile, and very hard.

Martensite is the first stage in the transformation of austenite. The structure of martensite is needlelike, and is much harder than austenite. Martensite is not very stable. The iron is a mixture of the alpha and beta forms.

Troostite is the next stage following martensite. Troostite is not stable. It is not quite so hard as martensite, probably because the proportion of alpha iron has been increased.

Sorbite is the last stage before pearlite. Sorbite is softer than troostite and harder and much stronger than pearlite.

Pearlite is the last stage in the transformation of austenite. Pearlite is a mixture of 6 parts of ferrite and 1 part of cementite, and it is less hard, less strong, and more ductile than the other forms of austenite. Pearlite is quite stable below Ar_1, except that the cementite tends to form ferrite and graphite.

306. Rapid Cooling of Molten Steel.—In the rapid cooling of molten steel the molecules do not have time to go through all of the successive transformations that they do in the case of slow cooling. Hence, various arrangements of the constituents will be found in the final structure of the steel due to the rate of cooling, initial temperature, amount of carbon, and amounts of other elements present. Hypo-eutectoid steel will always have free ferrite present with the other constituents, while free cementite will be found in the case of hyper-eutectoid steel.

Rapid cooling or quenching (as in water) of a solution of molten steel will give a steel whose constitution will be austenite with some martensite. Very rapid cooling, as by quenching in ice water, from the temperature $T1$, at which austenite begins to form, will give the same results.

A little slower rate of cooling from a molten steel solution, or a rapid cooling from the temperature $T2$ (of the formation of martensite) gives a steel consisting mostly of martensite with some austenite or troostite or both.

A slower rate of cooling (as quenching in air) from a molten steel solution, or a rapid cooling from the temperature $T3$, of the formation of troostite, will give a steel consisting essentially of troostite with some martensite but with no sorbite.

A very slow cooling of a molten steel solution, or a rapid cooling from the temperature $T4$, of the formation of sorbite, gives sorbite steel which may contain some pearlite.

If slow cooling is carried to the temperature $T5$ (of the formation of pearlite), the constitution of the resulting steel will be mainly pearlite, except that some sorbite will be present if the steel is suddenly cooled from a temperature very close to Ar_1.

Temperatures $T1$, $T2$, $T3$, and $T4$ are all above, while the temperature $T5$ is slightly below, the point Ar_1.

307. An Explanation of the Hardening of Steel.—It is thought that the hardness of steel and iron depends upon the amount of beta iron present, and that any treatment which will increase the amount of beta iron present will also increase the hardness. An increase in the carbon content increases the hardness probably because the carbon tends to hold the iron in the beta form and prevent its transformation into the softer alpha iron. Sudden cooling, from the temperature of the formation of beta iron, does not give the beta iron time for all of it to change into the alpha form; and the quicker the cooling, the less the change.

308. An Explanation of the Tempering of Steel.—In ordinary commercial tempering, the steel is heated to a temperature varying from 425 to 600 degrees Fahrenheit, and then quenched.

The quenching is done to prevent the temperature from increasing above the desired degree.

The transformation of one constituent to another depends upon the temperature reached, the rate of increase in temperature, and the length of time that the steel is held at that temperature. In heating cold steel, the austenite will be all changed to martensite when 400 degrees Fahrenheit is reached, and the martensite will be changed to troostite before 750 degrees Fahrenheit is reached. Further heating changes the troostite to sorbite, sorbite alone existing above 1,100 degrees Fahrenheit. Consequently, in commercial tempering, the resultant steel will be troostite, with some ferrite or cementite, depending on the carbon content. Only austenite, martensite, austen-troostite, and marten-troostite steels can be tempered, the commercial process having no effect on troostite or trooso-sorbite steels.

309. An Explanation of the Annealing of Steel.—In annealing, the steel is heated to higher temperatures than in tempering and is allowed to cool very slowly. This gives a steel which may be essentially sorbite or pearlite (though some troostite may be present if the annealing temperature is low) which is softer and more ductile, though less strong, than the martensite, austenite, or troostite. Ordinary annealing temperatures vary from 400 to 900 degrees Fahrenheit.

To reduce the coarse crystallization in the steel, the annealing temperature must be above the critical point, say from 1,700 to 1,450 degrees Fahrenheit, depending upon the carbon content. Some think that in annealing, most all of the beta iron present is changed over into the alpha form.

F. PHYSICAL AND MECHANICAL PROPERTIES AND USES OF STEEL

310. General.—The physical and mechanical properties of steel depend upon the methods of manufacture, mechanical working, heat treatment, and chemical composition.

The principal chemical elements affecting the properties of steel are carbon, silicon, sulphur, phosphorus, and manganese. Other chemical elements are present, but usually in such small quantities that they have practically no effect on the properties of the steel. Some chemical elements, such as nickel, chromium, copper, vanadium, etc., when present in appreciable quantities, also have their effect on the properties of steel. The effect of these elements will be considered in the paragraphs on special and alloy steels.

311. Effect of Carbon.—Carbon has a very great effect on the physical properties of steel, depending on the amount present and

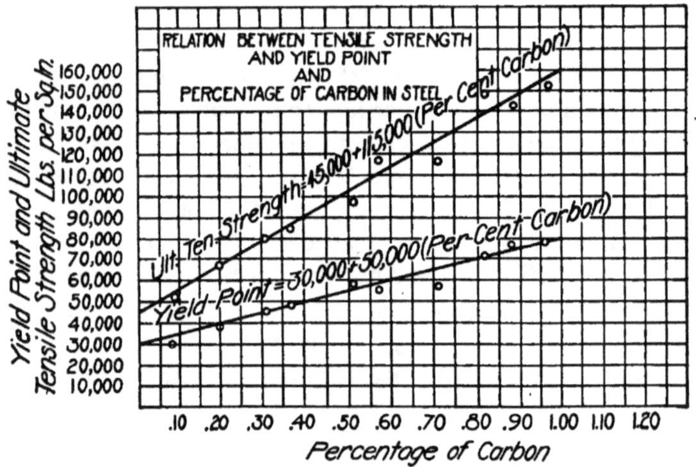

FIG. 106.—Effect of carbon on tensile strength of steel.

also upon its chemical composition with the steel. Carbon makes the steel hard and strong, but decreases the ductility. Steel may be classified according to the amount of carbon in it.

CLASSIFICATION OF CARBON STEELS

Kind	Per cent of carbon	Characteristics
Soft steel	0.10 to 0.20	Not temperable. Easy to weld.
Medium steel	0.20 to 0.40	Hard to temper. Weldable.
Hard steel	0.40 to 0.70	Temperable. Hard to weld.
Very hard steel	0.70 to 1.20	Easy to temper. Not weldable.

The following formulas are some approximate ones showing the relation of the carbon content to the tensile strength and ductility of steel:

Average steel,

Yield point = 30,000 + 50,000 × per cent C.

Ultimate tensile strength = 45,000 + 105,000 × per cent C.

Per cent elongation in 8 in. = 58 − (Ultimate tensile strength ÷ 2,500).

Per cent elongation in 8 in. = 1,500,000 ÷ ultimate tensile strength.

The last formula is one proposed by the American Society of Civil Engineers and has been used in many specifications for steel. It gives elongations that are too great for hard and very hard carbon steels, but it is fairly accurate for medium and soft steels.

Acid open-hearth steel,

Ultimate tensile strength = 45,000 + (108,000 × per cent C).

Basic open-hearth steel,

Ultimate tensile strength = 45,000 + (90,000 × per cent C).

If the percentages of phosphorus and manganese are known, then,

Acid open-hearth steel,

Ultimate tensile strength = 40,000 + (68,000 × per cent C) + (100,000 × per cent P) + (80,000 × per cent CM).

Basic open-hearth steel,

Ultimate tensile strength = 38,800 + (65,000 × per cent C) + (100,000 × per cent P) + (9,000 × per cent M) + (40,000 × per cent CM).

312. Effect of **Silicon, Sulphur,** Phosphorus **and Manganese.** *Silicon.*—The amount of silicon that is usually present in steel is less than 0.25 per cent, and up to this amount it has practically no effect on the steel. A larger amount, say 0.3 or 0.4 per cent increases the hardness, yield point, and ultimate strength with practically no reduction in the ductility. If the steel contains more than 1.0 per cent of silicon, it is called silicon steel.

Sulphur.—Within the limits common to ordinary steel (0.02 to 0.10 per cent), sulphur has practically no effect on the strength and ductility of cold steel, but it does have an injurious effect on the properties of hot steel. It causes what is known as "red shortness" (brittleness at a red heat) and makes the steel difficult to roll and weld. Larger amounts of sulphur tend to decrease the strength and ductility. The sulphur content is rarely over 0.06 per cent in good steel.

Phosphorus.—In small amounts phosphorus increases the strength very slightly, but it is a very undesirable element as it makes the steel very brittle and unable to resist shocks or blows. The phosphorus content rarely exceeds 0.07 per cent in good steel.

Manganese.—This element, in small amounts, increases the strength slightly and the hardness and malleability to a greater degree. The effect of manganese on steel increases with the amount of carbon present, and it has a greater effect on acid than on basic steel. From 2.0 to 6.0 per cent of manganese makes the steel very brittle. The amount of manganese present usually varies from 0.04 to 1.1 per cent, depending on the kind of steel. Steel containing over 6.0 per cent of manganese is called manganese steel.

313. Effect of Mechanical Working and Heat Treatment.—The effect of mechanical working on hot steel is to lessen the effect of large crystals, flaws, blowholes, etc., and to increase the solidarity, specific gravity, and strength. If the finishing temperature is above a red heat, the crystals may increase in size in cooling and the elastic limit and ultimate strength may be reduced. However, if the working is continued so that the finishing temperature is below a red heat, large crystals will be unable to form and the elastic limit and ultimate strength will be increased.

Cold working may be done on soft or medium steel, but not on hard steel as it is too brittle. The effect of cold working is to elongate the crystals, decrease the ductility, increase the brittleness, raise the elastic limit considerably, and raise the ultimate strength to a lesser degree.

Hardening makes the steel strong, hard, and brittle, while tempering removes some of the brittleness and increases the ductility and toughness without lessening the strength very much.

In general, annealing removes the effects of overstrain, increases the ductility, and lowers the strength. The effect of annealing depends in a large measure on the annealing temperature and the amount of carbon in the steel.

314. Tensile Strength of Steel.—The elastic limit in tension, which is about 50 or 60 per cent of the ultimate strength, varies from about 25,000 to 120,000 lb. per square inch, depending on the kind of steel. The yield point is usually from 3,000 to 5,000 lb. per square inch more than the elastic limit. The ultimate strength in tension varies from about 45,000 to over 200,000 lb. per square inch according to the kind of steel. The modulus of elasticity in tension is about 29,000,000 or 30,000,000 lb. per square inch, and is practically a constant for all kinds of steel.

The ultimate elongation in 8 in. usually ranges from 5 to 30 per cent, and the reduction in area from 10 to 60 per cent.

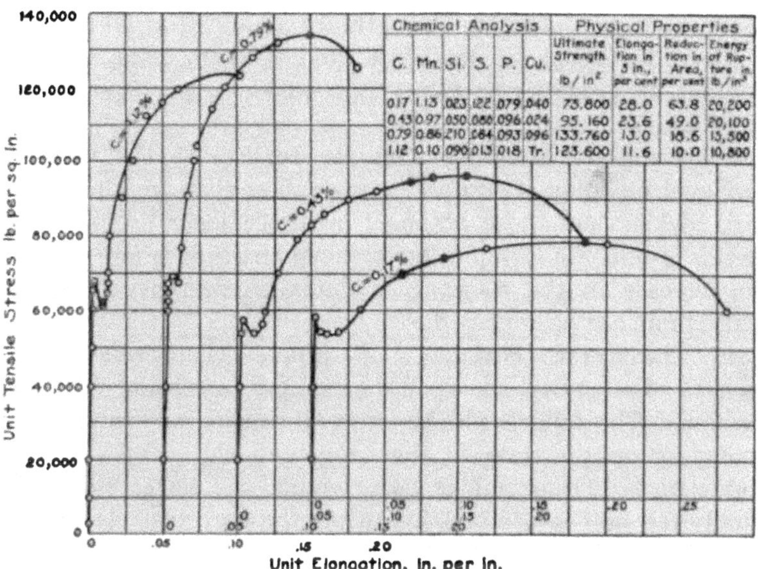

Chemical Analysis						Physical Properties			
C.	Mn.	Si.	S.	P.	Cu.	Ultimate Strength lb/in²	Elongation in 8 in., per cent	Reduction in Area, per cent	Energy of Rupture in lb/in³
0.17	1.13	.023	.122	.079	.040	73,800	28.0	63.8	20,200
0.43	0.97	.050	.080	.096	.024	95,160	23.6	49.0	20,100
0.79	0.86	.210	.084	.093	.096	133,760	13.0	18.6	13,500
1.12	0.10	.090	.013	.018	Tr.	123,600	11.6	10.0	10,800

FIG. 107.—Stress-strain diagrams for carbon steels in tension. (*Watertown Arsenal Tests.*)

The per cent elongation and the per cent reduction in area decrease as the strength increases, other things being equal.

The fracture of steel in tension is usually fine and silky in appearance, and varies from a full cup and cone for soft steel to a fracture that is square across for hard steel.

315. Compressive Strength of Steel.—The elastic limit and modulus of elasticity in compression are practically the same as in tension. Soft and medium steels have no ultimate strength in compression as the metal flows and flattens out as the load is increased, thus increasing the cross sectional area. The ultimate strength for structural steel is usually taken as 60,000 lb. per square inch.

Hard steel has a definite fracture in compression, the metal shearing off at an angle with the load axis. Sometimes the specimen breaks in many pieces. The strength in compression for hard steel is from 10 to 25 per cent greater than that in tension.

The tensile and compressive properties of steel are very closely related, and the two stress strain curves are very much alike in appearance.

316. Shearing Strength of Steel.—In shear, the values of the elastic limit and ultimate strength are about 70 or 80 per cent of those in tension for the same kind of steel. The shearing modulus of elasticity is about 12,000,000 lb. per square inch for all kinds of steel.

The computed shearing stress in torsion at the ultimate (torsional modulus of rupture) gives values that are about one-third higher than those obtained in direct shear. The reason for this is that the formula for torsional stress does not consider the decrease in the shearing modulus of elasticity after the elastic limit has been passed.

317. Transverse Strength of Steel.—The cross-bending strength of steel depends upon its strength in tension and compression. The failure will be either in tension or compression, depending on the form of the beam and the position of the neutral axis. Other things being equal, the softer steels will usually fail first in compression, while the harder steels will fail in tension. The elastic limit and modulus of elasticity have the same values in cross bending as in tension, while the ultimate is the same as that in tension or compression, depending on the kind of failure.

The deflection, up to the elastic limit, depends on the manner of loading, kind of supports, length of span, and the moment of inertia of the cross section (the modulus of elasticity being considered a constant for all steels). The maximum deflection at the time of failure is also dependent on the yield point and the ductility of the steel.

318. Average Properties of Rolled Carbon Steels.—The following table gives average properties of various rolled carbon steels. All stresses are in pounds per square inch, and elongations and reductions of areas are in percentages.

AVERAGE PROPERTIES OF VARIOUS ROLLED CARBON STEELS

Per cent of carbon	0.10	0.20	0.40	0.60	0.80	1.00
Elastic limit, tension and compression	30,000	35,000	50,000	60,000	70,000	80,000
Elastic limit, shear	18,000	21,000	30,000	36,000	42,000	48,000
Ultimate strength, tension	50,000	60,000	90,000	115,000	135,000	150,000
Ultimate strength, compression	(50,000)	(60,000)	105,000	135,000	155,000	175,000
Ultimate strength, shear	35,000	45,000	65,000	85,000	100,000	115,000
Modulus of rupture, cross bending	50,000	60,000	95,000	125,000	145,000	160,000
Modulus of elasticity, tension, compression, and cross bending	29,000,000 to 30,000,000					
Modulus of elasticity, shear	About 12,000,000					
Per cent elongation in 2 in	45	35	25	18	12	8
Per cent elongation in 8 in	30	25	18	14	10	6
Per cent reduction in area (tension)	60	50	37	27	20	15

Hard steels, which have been subjected to heat treatment, will probably have greater strength and less ductility than rolled hard steels. The process of manufacture, methods of treatment in rolling, and the presence of various chemical elements all have some influence on the strength and properties. Therefore, results from any individual test may differ considerably from the average values given in the table.

319. Effect of Combined Stresses.—In general, combined stresses tend to lower the elastic limit and ultimate strength in either kind of stress. Torsional stress will lower the elastic limit and ultimate strength in tension, compression, and cross bending, the amount of lowering being dependent on the amount of torsional stress present. Similarly, the elastic limit and ultimate in shear are lowered by the existence of any of the other stresses, depending upon their amount. This decrease or lowering of strength under any one kind of stress in combined stresses may be from 10 to 80 per cent, depending on the amount and kind of the other stress or stresses present.

320. Resistance to Impact Loads.—The resistance of steel to impact is of importance because steels are often subjected to blows and sudden loads in structural and other work. There is no generally accepted method for making impact tests; the test most commonly used being a drop test in which a given weight is dropped from a given height on a specimen of a certain size supported over a definite span on a heavy anvil block. Probably the best method of measuring the resistance of steel to impact is to determine the amount of work in impact required to deform a unit volume of the steel to rupture. This unit may be expressed in inch pounds per cubic inch or in foot-pounds per cubic inch. In a static test, the work required to deform the specimen to rupture is equal to the area under the stress strain curve. The values obtained from impact tests are usually from 20 to 30 per cent higher than those obtained from static tests.

The following table gives the results from some experiments made by Professor Hatt of Purdue University. The static tests were tensile tests.

Material	Tensile strength, pounds per square inch	Work of rupture in foot-pounds per cubic inch	
		Static	Impact
Soft steel	63,000	1,376	1,358
Boiler steel	80,000	1,230	1,855
Soft steel casting	62,000	1,450	2,315
Nickel steel	85,000	1,414	1,821
Locomotive-tire steel	140,000	976	2,918
Annealed wire	83,000	566	544
Steel wire	109,000	348	626

321. Ductility of Steel.—The ductility of steel may be measured by the elongation and reduction in area in a tension test (see Art. 318 for average values for rolled carbon steels).

The ductility of steel is often measured by a cold-bending test, which consists of bending a specimen over a sharp edge or about a pin or template of a certain radius. Cold bending may be done in a vise, on an anvil, or by the use of a steam hammer or a hydraulic press. Testing machines have been made especially for this test. The cold steel should be able to bend around a certain pin through an angle of the required number of degrees without cracking. The following table gives the values that ordinary carbon steel should bend without cracking.

Cold-bending Test of Steel

Kind of steel	Degrees of bend	Diameter of pin
Low-carbon steel	180	Zero
Medium-carbon steel	180	2 × the thickness of steel
High-carbon steel	90	3 × the thickness of steel
Very high-carbon steel	60	3 × the thickness of steel

322. Hardness of Steel.—Hardness may be said to mean any one of many things such as the ability of a tool to keep its cutting edge, the resistance of wheels and rails to wear due to rolling friction, resistance to abrasion, resistance to indentation, etc. There are many ways of measuring the hardness of steels, but only a few of the most common ones will be mentioned.

The Brinell method determines the resistance to indentation by measuring the indentation of a hardened sphere of a given size and under a given pressure. Results of tests made with a Brinell machine tend to show that the hardness of steel is directly proportional to its tensile strength. The Brinell method seems to be the best test yet proposed.

In the Shore scleroscope test, the hardness is determined by measuring the rebound of a pointed hammer that falls upon the metal from a definite height through a guiding glass tube. This method seems to give parallel results with the Brinell method.

In the Bauer drill test, a special drill is driven at constant speed under a fixed pressure. The hardness is measured by the depth of hole drilled in a given number of revolutions. The softer the steel, the deeper the hole.

323. Effect of Repeated and Alternating Stresses.—It has been found that a stress which is often repeated or alternated in a steel bar will finally cause a molecular change in the steel which lowers its strength and which may result in rupture of the bar. The repeated or alternating stress tends to break down the structure of the individual crystals and to cause them to fail by shear slips. This loss of strength is called "fatigue" of the steel.

The average stress causing the rupture of a steel bar depends upon the maximum stress applied to the extreme crystals, the behavior of these crystals, and the effect of the failure of a few crystals on the others. Small crystals resist repeated stresses better than large crystals do.

Results of tests have shown that steel may be ruptured by repeated applications of a stress much less than the ultimate, and that the number of repetitions required for failure increases very rapidly as the stress decreases. If the maximum stress is greater than the elastic limit, failure may occur after a comparatively few applications; but if the stress is much less than the elastic limit, an enormous number (millions) of repetitions will be required to cause failure. In general, the resistance of high-carbon steels to repeated stresses appears to be greater than that of low-carbon steels.

324. Welding of Steel.—Soft steel is quite easy to weld, medium steel is weldable, hard steel is very difficult to weld, and it is practically impossible to weld very hard steel. In soft steel the efficiency of the welded joint rarely exceeds 70 per cent, while in medium steel the efficiency is rarely over 55 per cent, even though the welding is done by expert workmen.

The thermit process for welding fractures in wrought iron and steel depends on the affinity of powdered aluminum for iron oxide. If these two substances are finely mixed and ignited, the resulting chemical action will raise the temperature of the adjacent metal to about 5,400 degrees Fahrenheit, which is high enough to allow of the melting and welding of any ordinary casting or forging. This process is much used for repairing breaks in large castings.

If acetylene gas and oxygen are properly mixed and ignited, they will generate a temperature of about 6,000 degress Fahrenheit which may be used to weld or fuse metals. The oxygen-acetylene process is much used for cutting metals.

In electric welding, an alternating-current dynamo of fairly high voltage is connected to a transformer whose secondary terminals give an electric current of large amperage and very low voltage (about 3 volts). These secondary terminals are connected to the welding clamps. The flow of a large current through the resistance offered by the two pieces of steel where they are in contact with each other causes the contact surfaces to be heated hot enough to weld or fuse together.

325. Magnetic Properties of Steel.—The magnetic properties of steel depend to a large extent on its composition, mechanical working, and heat treatment.

The carbon content should be low, less than 0.1 per cent.

Silicon decreases the hysteresis loss, increases the resistivity, reduces eddy currents, increases the permeability in a weak field, and decreases the permeability in a strong field.

More than 0.3 per cent of manganese appears to have a bad effect on the magnetic properties.

Sulphur and phosphorus have a bad effect on the magnetic properties of steel. The total sulphur and phosphorus content should be less than 0.15 per cent, and there should not be more than 0.10 per cent of either sulphur or phosphorus.

In general, hot steel (between 300 and 600 degrees Centigrade) has a slightly greater intensity of induced magnetism than cold steel. At about 200 degrees Centigrade, there seems to be a characteristic decrease in the intensity. Also, as the steel approaches closer to the critical point (700 to 800 degrees Centigrade) in temperature, the intensity decreases and becomes practically zero when the critical point is reached.

Cold working of steel appears to be somewhat injurious to the magnetism, but proper annealing seems to remove most of the bad effects. Annealing makes the steel softer and weaker and increases the magnetic properties. Hardening by quenching in oil or water has a bad effect on the magnetic properties. Tempering after hardening has a good effect in many cases.

In general, the weaker and softer a steel is, the more magnetic it is.

326. Specific Gravity and Coefficient of Expansion of Steel.—The specific gravity of steel is about 7.8, and the weight is about 490 lb. per cubic foot. Thorough mechanical working tends to increase slightly the specific gravity and weight per cubic foot.

The coefficient of expansion of steel is approximately 0.0000065 per degree Fahrenheit.

327. Summarized Specifications for Various Steels.—The following table gives the summarized requirements or specifications for various steels. Most of the values have been taken from the 1916 "Book of Standard Specifications of the American Society for Testing Materials." With some exceptions, the open-hearth process of manufacture is required. The nickel steel in the table should contain 3.25 per cent or more of nickel.

MINIMUM REQUIREMENTS FOR VARIOUS STEELS

Use	Carbon, per cent	Elastic limit, pounds per square inch	Ultimate tensile strength, pounds per square inch	Elongation in		Cold-bend test
				8 in.	2 in.	
Structural-rivet steel	0.08 to 0.15	30,000	55,000	30	..	Yes
Structural steel	0.18 to 0.30	35,000	60,000	25	..	Yes
Boiler-rivet steel	0.08 to 0.15	25,000	50,000	30	..	Yes
Boiler-plate steel	0.08 to 0.15	30,000	60,000	27	..	Yes
Boiler-flange steel	0.12 to 0.25	30,000	60,000	25	..	Yes
Boiler-firebox steel	0.12 to 0.25	28,000	57,000	27	..	Yes
Structural nickel steel	0.20 to 0.45	52,000	95,000	20	..	Yes
Nickel steel for rivets	0.15 to 0.30	45,000	75,000	20	..	Yes
Reinforced concrete bars	0.20 to 0.40	40,000	70,000	20	..	Yes
(Rerolled from rails)	0.20 to 0.50	50,000	80,000	15	..	Yes
Mild forge steel	0.10 to 0.20	30,000	55,000	..	25	No
Medium forge steel	0.20 to 0.40	40,000	75,000	..	20	No
Heat-treated forge steel	0.15 to 0.40	65,000	110,000	..	20	No
Machinery steel	0.35 to 0.60	40,000	75,000	20	..	
Rail steel	0.35 to 0.55	45,000	80,000	18	..	Yes
Steel railway tires	0.50 to 0.85	60,000	115,000	..	10	No
Axle steel	0.35 to 0.55	55,000	100,000	15	18	Yes
Gun steel	0.20 to 0.50	50,000	90,000	16	..	Yes
Spring steel	1.00 to 1.50	60,000	125,000	12	..	Yes
Tool steel	0.90 to 1.50	80,000	150,000	8	..	No
Cable-wire steel	0.70 to 1.50	100,000	200,000	6	..	..

328. Working Stresses for Structural Steel.—The allowable unit working stresses for steel depend on the kind of steel and the kind of loading. The working unit stress should never exceed the elastic limit of the steel. Often the allowable working unit stress for variable loads is taken at about half of the elastic limit. This value may be increased about 33 per cent for steady stresses; decreased about 30 or 40 per cent for repeated or alternating stresses; and decreased about 50 per cent for impact, shocks, or sudden loads. A simple rule for determining the allowable unit stress for varying loads is: Add to the maximum load the difference between the maximum and minimum loads and treat the sum as a static load. Then use the allowable working unit stress for steady stresses and design accordingly.

The following table gives average values of working unit stresses for structural steel under various loads:

Average Values of Allowable Working Unit Stresses for Ordinary Structural Steel

Stress	Material	Unit working stress in pounds per square inch			
		Variable	Steady	Alter-nating	Impact
Tension.......	Medium rolled or cast steel..... ..	16,000	21,000	10,000	8,000
Compression (short blocks)	Medium rolled or cast steel.......	16,000	21,000	10,000	8,000
Bending.......	Rolled beams and riveted beams..	16,000	21,000	10,000	8,000
Bending.......	Rolled pins, rivets, and bolts.....	20,000	26,000	13,000	10,000
Shear.........	Steel web plates, shop rivets, and pins..........................	10,000	13,000	6,500	5,000
Shear.........	Field rivets and pins.	9,000	12,000	6,000	4,500
Shear.........	Field bolts.......................	8,000	11,000	5,000	4,000
Bearing........	Shop rivets and pins.............	20,000	26,000	13,000	10,000
Bearing........	Field rivets......................	18,000	24,000	12,000	9,000

329. Uses of Steel.—Steel is probably the most important structural material in the world at the present time and its uses are practically innumerable. Some of its more important uses are: structural steel for buildings, ships, bridges, viaducts, and trestles; steel machinery of all kinds; railroad locomotives and cars; street railway cars; automobiles and wagons; dynamos, motors, and other electrical machinery; standpipes and tanks; hydraulic machinery; steel boilers and engines; guns, armor, and projectiles; tools, saws, lathes, planers, millers, and all kinds of metal and wood working machinery; cutlery, surgical instruments, and delicate apparatus of all kinds; wire; concrete reinforcement; etc.

The following are the uses of various carbon steels. The uses of the special and alloy steels will be given in the articles describing those steels.

USES OF VARIOUS CARBON STEELS

Carbon 0.05 to 0.15 per cent

Electrical sheet steel; boiler plate; rivets; bolts; stock for case hardening; nails; ship plate; forge work; sheet steel for tinning and galvanizing.

Carbon 0.15 to 0.25 per cent

General structural steel for bridges, buildings, etc.; forge steel; flange and firebox steel; ship plate; rivets; cold-rolled shafting.

Carbon 0.25 to 0.40 per cent

Machine parts in general; axles; shafts; connecting rods; piston rods; steel castings; reinforced concrete bars; medium forge steel.

Carbon 0.40 to 0.75 per cent

Railroad rails; steel castings; machinery steel; steel railway tires.

Carbon 0.60 to 0.80 per cent

Steel railway tires; hammers; cutlery; wood working tools; small lathe tools; small dies; cold sets; drills; taps; reamers; cold chisels; wire.

Carbon 0.80 to 1.00 per cent

Carbon steel springs; ordinary lathe tools; large dies, drills, and chisels; steel wire and cable.

Carbon 1.00 to 1.20 per cent

Large lathe tools; large and carefully heat-treated dies and drills; axes; hatchets; knives; spring steel; cable-wire steel.

Carbon 1.20 to 1.50 per cent

Some tool steels; spring steels; saws; files; jeweler's dies; balls and ball bearings.

CHAPTER XIV

SPECIAL STEELS AND CORROSION OF IRON AND STEEL

A. STEEL CASTINGS

330. Definition and Uses of Steel Castings.—Steel castings are unforged or unrolled castings made of bessemer, open-hearth, crucible, or any other kind of steel.

Steel castings are used in many places in preference to cast iron, and are often cheaper than built-up sections of structural steel. Some of the uses of steel castings are: bedplates, stern posts, stems, hydraulic cylinders, shaft struts, hawse pipes, stern pipes, crossing frogs, parts of railway cars, pinions, gears, hammer dies, gun carriages, various machine parts, etc.

331. Founding of Steel Castings.—In general, the founding of steel castings is the same as that for cast iron, except that much greater care must be used throughout.

The proper design of the patterns and cores is very important in steel castings. Intricate castings must be designed so that they will cool uniformly, or severe internal stresses or cracks will be formed. A large shrinkage must be allowed for, varying from 3⁄16 to 1⁄4 in. per foot. An uneven shrinkage is bad.

A special molding sand, consisting essentially of certain kinds of silica sand, must be used to prevent the steel from sticking to the molding material.

The molten steel used should have the proper chemical proportions to give good castings. Sufficient quantities of silicon and manganese should be present to eliminate blowholes and flaws, and to aid in the solidifying of the steel, though large quantities of these elements will cause brittleness. It is better to pour the metal very hot, as it makes smoother castings, and make allowance for the increased shrinkage. The castings are removed from the molds and cleaned in the same manner as iron castings.

Annealing a steel casting reduces the effect of overstrains caused by shrinkage and uneven cooling, has but little effect on the tensile strength, and increases the ductility, especially of medium and hard castings.

332. Properties of Steel Castings.—The physical qualities of soft, medium, and hard steel castings are about the same as the physical qualities of soft, medium, and hard steel, except that the elastic limit, elongation, and reduction of area are a little less. The effect of various chemical elements, heat treatments, etc. is the same on steel castings as on the ordinary steels. The chemical elements and treatments can be controlled so as to produce the steel castings that are the most suitable for the given purpose.

Good steel castings should rarely contain over 0.40 per cent of carbon, unless it is desired to secure a hard, strong casting that will resist wear. The phosphorus should be less than 0.07 per cent, and the sulphur less than 0.05 per cent.

The following table gives the minimum requirements for the strength and ductility of steel castings as specified by the American Society for Testing Materials:

MINIMUM REQUIREMENTS FOR STEEL CASTINGS

Physical quality	Soft castings	Medium castings	Hard castings
Ultimate tensile strength, lb. per square inch	60,000	70,000	85,000
Yield point in tension, lb. per square inch	27,000	31,500	38,250
Per cent elongation in 2 in.	22	18	15
Per cent reduction of area	30	25	20
Cold-bending test	Yes	Yes	No

B. ALLOY STEELS

333. Definition and Classification of Alloy Steels.—Alloy steels are those steels which owe their distinctive properties chiefly to the presence of an element or elements other than carbon, or jointly to the presence of some such element or elements and carbon.

Alloy steels are usually classified according to the presence of one or more principal elements besides carbon. Three part alloys are those whose properties are chiefly dependent on one other element besides carbon and iron; 4 part alloys contain two other such chief elements; and 5 part alloys have three other such elements. Tungsten steel, chromium-nickel steel, and chromium-nickel-vanadium steel are examples of 3, 4, and 5 part alloys respectively.

The names given to the various alloy steels indicate the names of the alloying elements present.

Only a brief summary of the properties and uses of some of the most important alloy steels will be given in the following paragraphs.

334. Heat Treatment of Alloy Steels.—The proper heat treatment of alloy steels is very important as it materially affects the strength and other properties. In the heat treatment the following things should be considered: the original chemical and physical properties of the metal; the composition of the gases and other substances that come in contact with the metal during the heating and cooling; the rate of increase of temperature; the maximum temperature reached; the length of time that the metal is kept at this temperature; and the rate of cooling.

In general, the heat treatment of alloy steels is easier to control than that of ordinary carbon steels because: (1) in many cases the critical temperatures are lower, and, consequently, the changes in structure at the critical temperatures are slower and more easily controlled; and (2) the substances in solution in the various forms of iron appear to retard and slow up the rate of change from one form to another.

Some elements, such as manganese and nickel, lower the critical temperatures. Other elements, such as chromium, lower the critical temperatures and also form special carbides; while some elements, of which tungsten, vanadium, molybdenum and probably aluminum, copper, and titanium are examples, form special carbides and only slightly lower the critical temperatures. Silicon acts in about the same way as carbon, and has practically no effect on the critical temperature. If a sufficient amount of silicon is present, it will throw the carbon out of solution as graphite and thus cause the steel to be too brittle for engineering use.

Frequently, a double quenching of an alloy steel gives very good results. The metal is first quenched to a certain temperature in one bath (such as a lead bath heated to the proper temperature) and then quenched again and completely cooled by being immersed in a second bath. Sometimes the metal is allowed to cool slowly in air after the first quenching.

In general, a high-grade alloy steel should be annealed after every process in the manufacturing, such as forging, pressing, rolling, rough machining, etc. The metal should also be annealed before hardening or case hardening.

335. Nickel Steel.—Nickel steel usually contains from 2.75 to 4.5 per cent of nickel and from 0.20 to 0.40 per cent of carbon. It combines hardness, great tensile strength, and high elastic limit with great ductility. The specific gravity is about the same as that of carbon steels, and increases slightly with an increase of nickel. The resistance to corrosion increases with increase of nickel up to about 18 per cent of nickel. Ordinary nickel steel, containing about 3.5 per cent of nickel, corrodes slightly less than carbon steel. Nickel steel has a high electrical resistance. Ordinary nickel steel possesses a magnetic permeability greater than wrought iron.

An addition of about 3.5 per cent of nickel to a medium carbon steel raises the elastic limit and ultimate about 50 per cent and increases the elongation about 15 per cent. Nickel steel has been made with elastic limits varying from 48,000 to 120,000 lb. per square inch; with ultimates varying from 90,000 to 277,000 lb. per square inch; and with elongations varying from 25 to 3 per cent in tension. Ordinary nickel steel rarely receives a heat treatment.

Nickel steel is used for guns, armor plate, structural steel, rivets, railroad steel, machine parts, axles, shafts, etc. Invar is a nickel steel (containing about 36 per cent of nickel) with a low-expansion coefficient.

336. Manganese Steel.—Manganese steel usually contains from 6 to 20 per cent of manganese with less than 1.5 per cent of carbon. The composition for the best results is about as follows: 12 to 15 per cent of manganese, less than 0.6 per cent of carbon, less than 0.4 per cent of sulphur, and less than 0.4 per cent of phosphorus. Manganese steel is strong, hard, tough, and malleable, and offers a high resistance to wear. This steel is usually cast to form, but can be hot worked at a yellow heat. It is softened by quenching from a yellow heat, and hardened by very slow cooling from high temperatures. Strength and ductility increase with manganese up to about 15 per cent, after which they decrease. The maximum tensile strength is more than 140,000 lb. per square inch, and the elastic limit more than 90,000 lb. per square inch. The maximum elongation is more than 40 per cent in 2 in., and the maximum reduction in area is over 50 per cent. The shrinkage of manganese steel is more than that of ordinary steel.

Manganese steel is used where hardness, toughness, strength, and malleability are desired as for rails, frogs, crossings, car wheels, axles, tires, and for parts of crushing, rolling, and grinding machinery.

337. Vanadium Steel.—Vanadium steel contains from 0.1 to 0.6 per cent of vanadium. This element gives the steel a very high elastic limit and ultimate strength without decreasing the ductility. This steel is also very tough and able to resist impact stresses when properly made. The maximum ultimate tensile strength varies from 60,000 to 180,000 lb. per square inch, and the elongation from 40 to 20 per cent, depending on the heat treatments and amounts of vanadium and carbon present.

Vanadium steel is used in many places where a high elastic limit and ultimate strength combined with a large shock resistance is desired, as in springs, axles, shafts, gears, parts of railroad rolling stock and automobiles, etc.

338. Chrome Steel.—Chrome steel contains from 1.0 to 2.5 per cent of chromium with from 0.8 to 2.0 per cent of carbon. The chromium increases the elastic limit, ultimate strength, and hardness. Chrome steel is used to some extent for cutlery as the chromium tends to make the steel rust proof. Though chrome steel has been used a little for structural work, it is rarely used now because of the discovery of better and cheaper alloy steels.

339. Silicon and Aluminum Steels.—Silicon steel resembles nickel steel in its properties, and is used mostly for the cores, pole pieces, etc. of electrical machinery. Silicon, up to about 4 per cent, increases the elastic limit and ultimate, after which there is a decrease. The ductility decreases as the silicon increases. For electrical work, about 3 per cent of silicon, with low percentages of the other elements, seems to give the best results.

Aluminum steel is no harder or stronger than ordinary steel, but it is more solid and has less blowholes. Aluminum tends to cause hot and cold shortness. The effect of aluminum is about the same as silicon; consequently, aluminum is rarely used as the silicon is much cheaper.

340. Tungsten, Molybdenum, and Cobalt Steels.—Tungsten steel contains from 3.0 to 5.0 per cent of tungsten together with a little chromium and manganese. The tensile properties resemble those of high-carbon steel, the ultimate strength and elastic limit being high, the ductility low, and the hardness quite great. This steel is used for tools in metal cutting, and some grades will cut when red hot. Tungsten is a hardener of steel.

Molybdenum steel contains from 0.3 to 3.0 per cent of molybdenum. This steel is similar to tungsten steel in its properties, 1 per cent of molybdenum having about the same effect as 2 or 3 per cent of tungsten. This steel is used for high-grade saws and a few other purposes. Molybdenum is more expensive than tungsten.

Cobalt steel has practically the same properties and uses as tungsten steel, and is coming into use in the making of high-speed tool steels.

341. Copper Steel.—Copper steel contains from 1.0 to 4.0 per cent of copper. Copper steel has about the same strength as nickel steel, but is a little more brittle and less ductile. Copper hardens and increases the strength of carbon steels without reducing their ductility. If the copper exceeds 4.0 per cent, the steel is very difficult to roll. Copper increases the electrical resistance. Copper steel costs less than nickel steel and has about the same uses.

342. Some Four and Five Part Alloys.—Chromium-nickel steel has great strength and hardness. Annealing rolled chromium-nickel steel reduces the elastic limit and ultimate and increases the ductility, while hardening the steel greatly increases the elastic limit and ultimate and decreases the ductility. The elastic limit may vary from 50,000 to 150,000 lb. per square inch, the ultimate strength from 85,000 to 200,000 lb. per square inch, and the elongation in 2 in. from 35 to 2 per cent, depending upon the elements present and the heat treatment. Chromium-nickel steels are used in making automobiles, guns, armor, safes, machine parts, gears, axles, etc. In making armor plate, a plate of this steel is heated and then the surface is suddenly cooled with a cold brine spray, thus making the surface very hard and strong and leaving the interior more tough.

Chromium-vanadium steel has a very high strength when tempered properly. Heat treatments have about the same effect on this steel as on carbon steels, though the effect is more pronounced. The elastic limit may vary from 60,000 to 190,000 lb. per square inch and the ultimate strength from 90,000 to 210,000 lb. per square inch, depending on the amounts of the elements present and the heat·treatments. Chromium-vanadium steel is used for locomotive forgings, shafts, springs, tires, tubes, plates, wire, etc.

Nickel-vanadium steel contains about 0.20 per cent of carbon, from 2 to 12 per cent of nickel, and from 0.7 to 1.0 per cent of vanadium. More than 1.0 per cent of vanadium decreases the strength and increases the brittleness. The addition of vanadium to a nickel steel raises the ultimate tensile strength, elastic limit, toughness, and hardness with very little decrease in ductility. Tempering improves the qualities. The strength of this steel is about the same as that of chromium-vanadium steel.

Tungsten-chromium-vanadium steel usually contains from 15 to 20 per cent of tungsten, 3 to 5 per cent of chromium, 0.5 to 2.0 per cent of vanadium, 0.60 to 0.80 per cent of carbon, and very low percentages of silicon, sulphur, and phosphorus. When properly heat treated, this steel offers a very high resistance to wear at ordinary and fairly high temperatures (up to and including a red heat) besides having high strength and fair ductility. On account of these properties this steel, which is frequently called high-speed steel, is used for cutting tools on metal working machinery and in places where heat and wear are to be resisted.

C. CORROSION OF IRON AND STEEL

343. Definition of Corrosion.—Corrosion of iron and steel is the oxidation of the iron due to the formation of rust or the gradual changing of the iron to iron oxide. This rust will form rapidly upon the exposed surfaces of the iron or steel, so that the life of an unprotected steel structure is much less than that of one of stone or wood. Rust will not form in perfectly dry air. An analysis of rust taken from a bridge in Conway, Wales, gave 93.1 per cent of Fe_2O_3, 5.8 per cent of FeO, and 0.9 per cent of carbon dioxide.

Any unprotected surface of iron or steel, when exposed to the atmosphere, soon becomes covered with a thin layer of iron oxide, which tends to protect the surface from further rusting. This thin layer of rust does practically no injury to the metal. With prolonged exposure to moist air, this film of oxide becomes deeper and assumes a reddish-brown color. The deeper the film of rust, the more the metal is injured. Sometimes the corrosive action forms pits in the metal, which later tend to enlarge and become deep holes. This pitting action is very injurious and also difficult to stop after it has once started.

344. The Life of Iron and Steel under Corrosion.—Corrosion is promoted by moisture, carbon dioxide, smoke, sulphurous vapors, acid vapors, acid waters, salt waters, sewage, decaying vegetable or animal matter, and in general by all kinds of impurities, both in and adjacent to the steel. Electrolytic action and electrolysis also originate and hasten corrosion.

It is thought that structural steel rusts more rapidly than wrought or cast iron, but this has not been definitely proved. Overhead bracing of bridges exposed to the smoke of passing locomotives rusts rapidly. Old wrought-iron bridges often show less corrosion than do the modern steel ones. Metal that is subject to shocks and overstrain is more liable to corrosion than metal subject to steady stresses. Because they are more homogeneous, smooth surfaces rust less than rough surfaces. Poor steel containing blowholes and imperfections rusts rapidly for the same reason. Steel containing few impurities rusts less than steel containing many impurities.

The life of steel, wrought iron, and cast iron under corrosion depends on the amount of protection that has been given to the exposed surfaces, and also upon the condition of the air and water surrounding the metal. The following table gives an idea of the life of unprotected plates under certain conditions:

YEARS OF *LIFE* OF AN UNPROTECTED PLATE OF A CERTAIN SIZE

Material	Sea water		Fresh water		Air	
	Foul	Clear	Foul	Clear	Impure	Pure
Cast iron	15	16	26	88	21	88
Wrought iron	5	8	7	81	8	81
Steel	5	10	9	80	8	80
Cast iron, skin removed	4	11	14	90	12	90
Cast iron, galvanized	11	28	30	208	50	208

345. Theories of Corrosion.—The carbon-dioxide theory of corrosion is that carbon dioxide combines with the iron to form iron carbonate. Then, under the action of free oxygen in the air, this iron carbonate separates into iron oxide and carbon dioxide, thus furnishing a new supply of carbon dioxide so that the process can be continued. This theory does not give a complete explanation of corrosion as iron will often rust when there is no carbon dioxide present.

The moisture theory of corrosion is that water acts upon and combines with the iron producing iron oxide and hydrogen peroxide, but there are few proofs of this action. It is true, however, that water is an agency that promotes corrosion, especially when it contains acids or sulphurous compounds.

The electrolytic theory of corrosion is that the corrosion is caused by small momentary electric currents that originate at points in the metal where it is not homogeneous. Impurities, large grains, roughness of surface, non-uniformity of structure, etc. appear to cause a difference in potential in neighboring parts of the metal. This difference of potential causes small currents to flow, and these currents cause the oxidation of the iron. The electrolytic theory is generally accepted as the best theory yet proposed, even though the proof is not quite complete.

346. Prevention of Corrosion in General.—In general, corrosion may be prevented by excluding the water and the atmosphere from the surfaces of the metals by covering them with preservative coatings of oils, paints, varnishes, tar or bituminous compounds, concrete; or coating or plating the surfaces with aluminum, tin, nickel, lead, or zinc, or sometimes by the black oxide of iron. For surfaces properly protected, the figures in the above table for the life of metals may be doubled or tripled if the metal is not subject to shock or wear.

The resistance of steel to corrosion may be increased by taking certain precautions in the manufacture so as (1) to produce a metal containing very few or no impurities which will set up an electrolytic action with the iron; (2) to add some substance to the metal which will act to inhibit electrolysis of the iron; or (3) to roll the metal in special rolls so as to make a special fine grained surface that is very dense and mechanically resistant to corrosion.

347. Prevention of Corrosion by Painting.—Painting is the usual way of preserving iron and steel from corrosion. Some of the paints used are asphalt, chrome, white and red lead, white zinc, graphite, and iron oxide paints, as well as various other paints and varnishes some of which are supposed to contain rust repelling or resisting constituents.

The surfaces of the metal must be thoroughly cleaned. Mud, dirt, rust, scale, etc. can be removed with steel brushes, scrapers, or chisel and hammer. The use of a sand blast is the best method of cleaning metal surfaces before painting. Sometimes rust and scale are removed by pickling with a 10 per cent sulphuric acid solution. All traces of the acid must be removed by washing, and the metal thoroughly dried before painting.

At least two coats of paint should be applied, and three or four coats are better. Each coat of paint should be allowed to dry before the next is applied, the time between coats being at least a week. The paint should be applied in a smooth even coat (with special care being taken at the angles and edges) so as to form an impervious film.

One gallon of paint will usually cover from 250 to 400 sq. ft. as a first coat, depending on the surface conditions, and from 350 to 500 sq. ft. as a second coat.

348. Prevention of Corrosion by Covering with Concrete or Asphalt.—Cement mortar or concrete will protect the surfaces of iron and steel from corrosion when the metal is embedded in the mortar or concrete, and there are no cracks or flaws allowing the penetration of water or other corrosive agents. Oil or oil paints should not be used on iron and steel that are to be placed in concrete.

Asphalt will protect iron and steel from corrosion if it is applied in appreciable thickness and as a continuous coat. The asphalt should be slightly elastic, when cold, should have a high melting point, should not soften much at 100 degrees Fahrenheit, and should be applied at a temperature of 300 or 400 degrees Fahrenheit. The surface of the metal must be dry and should be hot, and the coating should be of considerable thickness. The usual ways of applying asphalt are either by dipping the metal in the asphalt or by pouring the asphalt on the metal.

349. Prevention of Corrosion by Galvanizing.—Galvanizing is the depositing of a thin coating of zinc on the surfaces of iron or steel. Zinc is probably the best preservative of iron and steel against corrosion. Besides forming a complete covering for the iron, zinc is highly electro-positive with respect to the iron, and the iron will not corrode when the surfaces of the two metals are in contact. When two metals are in contact with each other, one is electro-positive and the other is electro-negative. The electro-positive metal may corrode while the electro-negative metal will not corrode.

In ordinary, or hot, galvanizing, the articles are first cleaned by pickling and are then dipped in a hydrochloric acid solution before being immersed in a bath of molten zinc whose temperature is from 800 to 900 degrees Fahrenheit. When the articles have reached the temperature of the bath, they are withdrawn, and the coating is set in water. Wire, bands, and similar articles are drawn continuously through the bath.

Sheradizing is the heating of the iron or steel articles in the presence of powdered zinc, which forms a zinc coating on them. A closed retort is used and the temperature is below the melting point of the zinc. The time required varies from 30 minutes to several hours, depending on conditions.

The American standard test for galvanized wire is as follows: Prepare a neutral solution of sulphate of copper (specific gravity of 1.185). Dip the wire in this solution for 1 minute, wash, and wipe dry. To pass the test, the wire must stand 4 dips without showing a permanent coating of copper on any part of the wire.

350. Prevention of Corrosion by Aluminum, Nickel, Tin, or Lead Plating.—Aluminum, nickel, tin, or lead plating consists of depositing a thin coating of one of these metals on the surfaces of wrought iron, cast iron, or steel as a protection from corrosion. Good aluminum coatings are difficult to secure. Nickel plating makes a fine appearing surface which is capable of taking a high polish. Pinholes are often found in tin plating, through which the atmosphere can attack the iron and cause corrosion. A lead coating is a good protection for iron and steel, provided that the coating is perfectly gastight. Sheets, having a thin coating of lead, are known as terne plates and are not very durable.

351. Prevention of Corrosion by the Inoxidation Process.—This process is the forming of a continuous coating of black oxide of iron over the surfaces of the iron or steel. This coating is stable and affords good protection against corrosion. The usual method is to heat the articles to a temperature of 1,200 degrees Fahrenheit or more in a closed retort, and then add steam (sometimes partially decomposed steam) at a temperature of 1,200 degrees Fahrenheit or more and a hydrocarbon such as producer gas, naphtha, etc. This addition of steam and hydrocarbon forms a film of iron oxide over the metal. The time required varies from 15 minutes to 2 hours, depending on the details of the process used and other conditions.

CHAPTER XV

NON-FERROUS METALS AND THEIR ALLOYS

A. THE NON-FERROUS METALS

352. General.—The non-ferrous, or minor, metals may be divided into three classes according to their industrial importance. In the first class are aluminum, copper, lead, nickel, tin, and zinc, while the second class contains antimony, bismuth, cadmium, gold, silver, mercury, and platinum. The third class contains metals whose chief use is as an alloying element, such as cobalt, chromium, magnesium, manganese, molybdenum, titanium, tungsten, vanadium, etc.

353. Copper.—Copper is a reddish colored metal and, next to iron, is the most useful and valuable metal. It occurs in native form and in ores, usually as a sulphide or oxide. The most important ore is copper pyrites, $CuFeS_2$.

The methods of extracting copper from its ores vary greatly because of the presence of different impurities. In general, the process consists of (1) roasting to remove the sulphur; (2) smelting the ore in a blast furnace to remove most of the gangue as slag; and (3) the conversion of the blast furnace product, a mixture of metallic sulphides, called "matte," into copper by using a small type of bessemer converter. After this the copper is refined, usually by an electrolytic process, to make it purer.

Native copper is crushed, concentrated, smelted in a reverberatory furnace, and then refined.

Copper is malleable and ductile and may be cast, rolled or drawn, and annealed. It oxidizes at a red heat and melts at about 1,900 degrees Fahrenheit. Its specific gravity varies from about 8.6 in castings to about 8.9 in rolled pieces. Copper is non-corrodible in dry air. One of the most important properties of copper is its high electrical conductivity, which increases with its purity. The modulus of elasticity in tension is about 16,000,000 lb. per square inch. The following table gives average values of the tensile strength:

Tensile Strength of Copper

Kind	Elastic limit, pounds per square inch	Ultimate, pounds per square inch	Per cent elongation
Cast	8,000	25,000	
Rolled	15,000	40,000	
Soft wire	15,000	40,000	30
Medium wire	25,000	50,000	4
Hard wire	35,000	60,000	2½

More than half of the copper, including copper wire, is used in electrical work; about one-fourth in brass works; and the remainder in copper sheets, castings, alloys, etc.

354. Lead.—Lead is next to copper in commercial importance. It occurs usually as a sulphide mixed with some silver or antimony. The principal ore is galena, PbS. Lead is extracted from its ores by first roasting the ore and then smelting it in a blast furnace. Sometimes the silver or other important elements are removed. If the lead contains too many impurities, it is refined.

Lead is a very soft, non-corrodible (after the formation of an oxide film on the surface), plastic, inelastic, bluish white metal with a very low strength. Its specific gravity is about 11.3, and its melting point about 625 degrees Fahrenheit. The tensile strength of lead varies from 1,600 to 2,400 lb. per square inch. Lead flows under very small unit loads; is malleable and ductile; and can be drawn into wire.

Lead is used in alloys, plumbing work, paints, chemical manufacture, and for lead pipes, bullets, linings of tanks, chests, and pipes, etc.

355. Zinc.—Zinc is next to lead in order of importance. It occurs in the form of ores, the most important of which are zinc blende, ZnS, calamine or zinc spar, $ZnCO_3$, and hemimorphite or zinc silicate, Zn_2SiO_4. Zinc ores are usually crushed, coneentrated, calcined, and roasted, after which the zinc is distilled and condensed. This crude zinc, or "spelter," must be refined before using if there is too much lead or iron present.

Zinc is a hard, brittle, bluish white metal with a specific gravity of about 6.9 when cast and 7.1 when rolled. Its fracture is crystalline in appearance. The principal impurities present are lead, iron, and cadmium. The maximum total amount of these impurities should not exceed 0.10 per cent for high-grade spelter, 0.50 per cent for intermediate spelter, 1.20 per cent for brass special, with no limits for prime western and dross grades. Lead makes the spelter softer, but more than 0.7 per cent causes cracking. Iron and cadmium make the spelter harder and more brittle. Cadmium also tends to increase the cracking. Average values for strength are: 5,000 lb. per square inch for tension, 12,000 lb. per square inch for cross bending, and 20,000 lb. per square inch for 10 per cent compression.

Zinc is used for galvanizing and alloys principally, and also for zinc dust, zinc castings, sheet zinc, etc.

356. Tin.—Tin is a lustrous white metal occurring in nature in ores which may be found in lodes or veins or in alluvial deposits. The principal ore is cassiterite, SnO_2. Tin ore is first concentrated and then smelted, after which the crude tin is refined.

Tin is a malleable metal with a specific gravity of 7.3 and a melting point of 450 degrees Fahrenheit. Its strength is about 3,500 lb. per square inch in tension, 6,500 lb. per square inch in compression, and 4,000 lb. per square inch in cross bending.

Tin is used for tin plates, tin cans, safety plugs for boilers, household articles, tin plating, tin foil, etc.

357. Aluminum.—Aluminum is a soft, white, malleable metal that is found in nature in many combinations. The two most important combinations are bauxite (a mixture of aluminum and iron hydrates with impurities) and cryolite (a mixture of sodium and aluminum fluoride). The principal method of extraction of aluminum from its ores consists in the electrolysis of comparatively pure aluminum dissolved in a bath of molten cryolite.

The melting point of aluminum is 1,150 degrees Fahrenheit; the specific gravity is about 2.55 when cast and 2.75 when rolled. Aluminum may be obtained from 99 to 99.5 per cent pure, the most common impurities being iron and silicon. Aluminum is practically free from corrosion, can be annealed, is quite ductile, and shrinks greatly when cast. The electrical conductivity of aluminum is from ½ to ⅔ that of copper. The following table gives an idea of the average strength of aluminum:

AVERAGE STRENGTH OF ALUMINUM

Kind	Tension			Compression		Tension and compression modulus of elasticity, pounds per square inch
	Elastic limit, pounds per square inch	Ultimate, pounds per square inch	Per cent reduction of area	Elastic limit, pounds per square inch	Ultimate, pounds per square inch	
Cast........	8,000	20,000	15	4,000	12,500	9,000.000
Rolled.......	18,000	30,000	30	4,000	11,500	13,500,000
Wire drawn..	25,000	45,000	50	.....	......	17,000,000

Aluminum is used for electrical work, alloys, thermit welding, wire, cooking utensils and where a strong, light, malleable, and non-corrodible cast or rolled metal is desired.

358. Nickel.—Nickel is the least important of the metals of the first class and perhaps it should be included in the third class rather than the first. Its principal ores are nickel pyrites (nickel sulphide) and garnierite (nickel magnesium silicate). The sulphur ore is first roasted and then smelted in a blast furnace, after which it may be run through a converter to remove the iron or refined to remove the copper. The silica ores are first mixed with some sulphur compounds, smelted in a blast furnace, and then treated in the same ways as the sulphur ores.

It is a hard, tough, ductile, malleable, and non-corrodible metal of a silvery white color and is capable of taking a high polish. It has a specific gravity varying from 8.3 to 9.2, and a melting point of 3,000 degrees Fahrenheit.

Nickel is used very much in metal plating and in alloys with copper and steel.

359. Gold, Silver, and Platinum.—Gold is the most malleable and ductile of all metals. One ounce Troy can be beaten to cover 160 sq. ft. of surface and 1 grain can be drawn into a wire 500 ft. long. The specific gravity of pure pressed gold is about 19.3. The melting point is about 1,915 degrees Fahrenheit. Gold is used for coins, ornaments, plating, etc.

Silver is the whitest of all metals, is very malleable and ductile, and is between gold and copper in hardness. The specific gravity varies from 10 to 11. The melting point is about 1,750 degrees Fahrenheit. It is the best heat conductor of all metals, and is equal to copper as a conductor of electricity. Silver is used for coins, ornaments, plating, electrical work, etc.

Platinum is a malleable and very ductile metal of a whitish gray color. It is very difficult to fuse. Platinum is used for chemical laboratory vessels, in the manufacture of electric incandescent lamps and electric contact points, for standard weights and measures, for jewelry, etc.

360. Some Other Non-ferrous Metals.—Antimony is a brittle metal of a bluish-white color and a highly crystalline or laminated structure. It has a specific gravity of about 6.75. It melts at 842 degrees Fahrenheit, and burns in air with a bluish-white flame. Antimony is used in the manufacture of alloys such as type metals and unit friction metals.

Bismuth is a very brittle metal of a light-reddish color and a highly crystalline structure. Its specific gravity is about 9.8. It melts at 510 degrees Fahrenheit and boils at 2,300 degrees Fahrenheit. The conductivity for electricity is about $\frac{1}{80}$ that of silver. Bismuth is very diamagnetic. Its tensile strength is about 6,400 lb. per square inch. Bismuth is used in alloys.

Cadmium is a lustrous bluish-white metal with a fibrous fracture. Its specific gravity is about 8.65. It melts below 500 degrees Fahrenheit and volatilizes at about 680 degrees Fahrenheit. Cadmium is used in the manufacture of fusible alloys, etc.

Magnesium is a light, malleable, and ductile metal of brilliant silver-white color. It is non-corrosive. It is highly combustible and burns with a very brilliant light. The specific gravity is about 1.70. It melts at 1,200 degrees Fahrenheit. Magnesium is used for signal lights, flash lights, and as an alloy with aluminum.

Manganese has a specific gravity of about 7.5. It is used as an alloying material with iron, especially in the manufacture of steel, and also as an alloy with steel.

B. ALLOYS OF NON-FERROUS METALS

361. General.—An alloy is made by combining two or more metals, when in a molten condition, and this alloy may be either a solid solution of the metals or their chemical compounds, or a mixture of such solutions. The physical properties of an alloy cannot be predicted from the properties and proportions of the metals used.

362. Brasses.—Ordinary brasses are alloys of copper and zinc, while special brasses contain some other element or elements in addition. The most valuable brasses contain from 65 to 80 per cent of copper and from 35 to 20 per cent of zinc. A little tin, about 1 or 2 per cent, is usually added if the brass is to be turned or planed. These mixtures give a strong ductile alloy that can be cast, drawn, or rolled. Brass has a specific gravity of 8.95, can be annealed, and is harder and more ductile than copper. The tensile strength of brass is about 25,000 lb. per square inch, and this can be increased by rolling or drawing.

The addition of lead to brass softens it and lowers the strength and ductility. More than 3 per cent of lead should not be used because of the danger of segregation.

Aluminum, up to 5 per cent, increases the hardness, elastic limit, and ultimate in tension, but decreases the ductility. Aluminum brass is used for machine castings, forgings, plates, etc., and is a strong non-corrosive brass.

The most valuable special brass is the so-called "manganese bronze." The manganese is used for deoxidation purposes in the manufacture, and the resultant brass contains practically no manganese. The chemical composition is variable; one being given as about 59 per cent of copper, 40½ per cent of zinc, less than 0.5 per cent of aluminum, and less than 0.15 per cent of lead; while another is given as 57 per cent of copper, 40 per cent of zinc, 1 per cent of tin, 1½ per cent of iron, and a little aluminum. As the range of composition is large, the properties vary accordingly. A soft grade of manganese bronze may have a tensile strength of 60,000 lb. per square inch and an elongation of over 40 per cent in 2 in., while a hard grade may have a tensile strength of 90,000 lb. per square inch and an elongation of 20 per cent in 2 in. This metal has great strength and toughness and can be cast in intricate forms successfully besides being highly resistant to corrosion.

AVERAGE STRENGTH OF CAST AND ROLLED MANGANESE BRONZE

Kind	Tension, pounds per square inch		Per cent elongation in 2 in.	Compression, pounds per square inch	
	Elastic limit	Ultimate		Elastic limit	Ultimate
Cast	30,000	75,000	28	37,000	95,000
Rolled or forged	45,000	95,000	22	55,000	140,000

Sterro metal is brass containing about 1½ per cent of iron. The iron increases the strength and working qualities.

Delta metal is brass containing about 3 per cent of iron. When delta metal is cast, its tensile strength is about 45,000 lb. per square inch with an elongation of about 10 per cent in 2 in. When it is rolled, its strength and ductility are about the same as those of medium carbon steel. This alloy has a high resistance to corrosion.

Tobin bronze is similar to delta metal except that its iron content is less and that small amounts of tin and lead have been added. This alloy may be cast or rolled, and it has a high tensile strength and a very high resistance to corrosion.

APPROXIMATE STRENGTH OF COPPER ZINC ALLOYS

Copper, per cent	Zinc, per cent	Per cent elongation in 5 in.	Tensile elastic limit, pounds per square inch	Ultimate strength in pounds per square inch		
				Tension	Compression	Cross-bending
100	0	7.0	14,000	27,000	41,000	30,000
80	20	28.0	8,000	28,000	40,000	23,000
70	30	22.0	8,500	32,000	45,000	27,000
60	40	20.0	17,000	40,000	75,000	40,000
50	50	5.0	19,000	30,000	115,000	35,000
20	80	0.5	8,000	10,000	50,000	23,000
0	100	0.5	4,000	5,500	22,000	7,000

363. Bronzes.—Bronzes are alloys of copper and tin. They are harder, denser, and more fusible than copper. Practically all of the commercial bronzes contain less than 25 per cent of tin.

The addition of lead increases the plasticity of bronzes. Lead bronzes are used for bearings.

Phosphor-bronze is a bronze containing less than 1 per cent of phosphorus. The addition of the phosphorus permits the molding of very perfect castings. This bronze is hard and tough, and its tensile strength varies from 40,000 to 100,000 lb. per square inch. It is used for bearings, valve seats, telephone wire, etc.

APPROXIMATE STRENGTH OF COPPER TIN ALLOYS

Copper, per cent	Zinc, per cent	Per cent elonga-tion in 5 in.	Tensile elastic limit, pounds per square inch	Ultimate strength in pounds per square inch		
				Tension	Com-pression	Cross-bending
100	0	7.0	14,000	27,000	41,000	30,000
95	5	12.0	16,000	30,000	42,000	33,000
90	10	5.0	19,000	30,000	47,000	39,000
80	20	0.05	21,000	35,000	75,000	55,000
75	25	0.0	22,000	22,000	110,000	32,000
70	30	0.0	5,500	5,500	140,000	12,000
65	35	0.0	2,200	2,200	85,000	5,000
45	55	0.0	3,000	3,000	35,000	5,000
10	90	6.5	3,500	6,500	10,000	5,500
0	100	35.0	.....	3,500	6,500	4,000

364. Various Aluminum Alloys.—Aluminum bronze is not a bronze, but is a copper aluminum alloy containing from 88 to 95 per cent of copper and from 12 to 5 per cent of aluminum. This alloy is very strong and ductile, but it shrinks very much in casting. With 10 per cent of aluminum, rolled specimens have an elastic limit of 60,000 lb. per square inch, an ultimate tensile strength of about 100,000 lb. per square inch, and an elongation of 10 per cent. With from 5 to 7 per cent of aluminum, the alloy has about the same properties as medium steel and, in addition, is non-corrosive. The modulus of elasticity in tension is about 18,000,000 lb. per square inch.

Aluminum zinc alloys contain up to 33 per cent of zinc and are light, hard, strong, and easily made.

Aluminum copper alloys contain up to 8 per cent of copper. The copper raises the yield point and ultimate strength but greatly lowers the ductility.

Aluminum magnesium alloys, "magnalium," contain about 1½ per cent of magnesium and are lighter, stronger, and harder than aluminum. The specific gravity is 2.5. These alloys can be forged, rolled, drawn, machined, and filed. They take a high polish and are resistant to oxidation.

Aluminum-copper-zinc alloys contain from 9 to 27 per cent of zinc and from 3 to 5 per cent of copper. These alloys are strong but not ductile.

In aluminum-copper-tin alloys additions up to 7.5 per cent of copper and 7.5 per cent of tin strengthen the aluminum. Alloys containing from 20 to 60 per cent of any one of these three metals are very weak. An alloy of 7.5 per cent of copper, 7.5 per cent of tin, and 85 per cent of aluminum has a tensile strength of about 30,000 lb. per square inch, an elongation of 4 per cent, and a specific gravity of about 3.02.

365. Various Nickel Alloys.—Invar is an alloy containing about 36 per cent of nickel, 63 per cent of iron, 0.3 per cent of carbon, and 0.7 per cent of manganese. It has a very low coefficient of expansion (about 1/28 of that of steel) and is used for steel tapes and other measuring instruments.

German silver is composed of copper, nickel, and tin or zinc in varying proportions. The best varieties contain from 18 to 25 per cent of nickel, 25 to 30 per cent of tin, and 45 to 57 per cent of copper. Sometimes tin is used instead of zinc, producing an inferior alloy. German silver is used for ornaments, table wear, etc.

Monel metal contains about 72 per cent of nickel, 1.5 per cent of iron, and 26.5 per cent of copper. It is ductile, flexible, easily soldered, has a high resistance to corrosion, and is suitable for roofing. The ultimate tensile strength of castings is about 75,000 lb. per square inch, and of good rolled sheets about 100,000 lb. per square inch.

Constantin contains about 40 per cent of nickel and 60 per cent of copper. Its electrical resistance is from 28 to 30 times that of copper. This alloy is used for electrical resistance wire and also for thermo-couples.

Tableware alloy consists of 25 per cent of nickel, 25 per cent of zinc, and 50 per cent of copper.

366. Bearing Metal Alloys.—A good bearing metal should have the following five characteristics:

1. A hearing metal must be strong enough to carry the load without any distortion. Some loads are as great as 400 lb. per square inch.

2. A good bearing metal should not heat rapidly. In general, the harder the bearing metal, the more likely it is to heat.

3. A good bearing metal should work well in the foundry and not produce spongy castings.

4. A good bearing metal should show but little friction. While friction in bearings depends largely upon the lubricant used, the metal also has some influence.

5. Other things being equal, the best bearing metal is the one which wears the slowest.

The principal constituents of bearing metals are copper, tin, lead, zinc, antimony, and iron.

PROPORTIONS OF SOME BEARING METALS

Kind or name	Percentages of					
	Copper	Tin	Lead	Zinc	Antimony	Iron
Gun metal	90.0	8.0	....	2.0		
Babbitt metal	3.0	89.0	....	...	3.0	
Repair babbitt	....	....	80.0	...	20.0	
Bearing metal for heavy bearings	77.0	8.0	15.0			
Engine brass	76.5	11.75	....	11.75		
Heavy bearings	10.0	....	65.0	...	25.0	
Ordinary bearings	....	15.0	73.0	...	12.0	
White metal	5.0	85.0	....	...	10.0	
Car brass lining	....	....	85.0	...	15.0	
American anti-friction metal	....	....	78.5	1.0	19.75	0.75
Car-box metal	....	....	84.5	...	14.3	0.7
Anti-friction metal	....	....	88.0	...	12.0	
Ajax metal (arsenic or phosphorus 0.4 per cent)	81.25	11.0	7.25			
Alloy "B"	77.0	8.0	15.0			
"K" bronze	77.0	10.5	12.5			
Standard (phosphorus 0.8 per cent)	79.7	10.0	9.5			

367. Fusible Alloys.—The following table gives the proportions of some fusible alloys and their approximate melting temperatures. Some of the temperatures may be inaccurate. Fusible alloys are used for fusible plugs in boilers, etc.

FUSIBLE ALLOYS

PROPORTIONS	MELTING TEMPERATURE IN DEGREES FAHRENHEIT
1 lead, 1 tin, 1 bismuth, 1 cadmium	155
2 lead, 1 tin, 4 bismuth, 1 cadmium	165
1 lead, 1 tin, 1 or 2 bismuth	200
3 tin, 5 bismuth	205
1 lead, 4 tin, 5 bismuth	240
1 lead, 1 bismuth	257
1 tin, 1 bismuth	285
2 tin, 1 bismuth	336
8 tin, 1 bismuth	392
1 lead, 1 tin	370 to 400
2 lead, 1 tin	440 to 475

368. Solders.—Common solder contains about 1 part tin and 1 part lead; fine solder, hard solder, and plumbers' solder contain 2 parts tin and 1 part lead; and cheap solder contains 1 part tin and 2 parts lead.

Common pewter contains 4 parts lead and 1 part tin.

Ordinary gold solder contains 14 parts gold, 6 parts silver, and 1 part copper.

Silver solder contains 145 parts silver, 73 parts brass (3 copper and 1 zinc), and 4 parts zinc.

German silver solder contains 38 parts copper, 54 parts zinc, and 8 parts nickel.

Aluminum solder contains 100 parts tin and 5 parts lead or zinc.

369. Composition and Use of Some Miscellaneous Alloys.—

Name	Copper	Zinc	Tin	Lead	Anti-mony	Bis-muth	Use
Type metal	..	..	5	75	20	..	Printing
Hard type metal	..	75	..	..	25	..	Printing
Bell metal	75	..	25	..	..	..	Bells
Best valve metal	86	2	12	..	..	..	Valves
Valve metal	83	15	2	..	..	..	Valves
Tobin bronze	60	38	1½	½	..	..	Various uses
Naval brass	62	37	1	..	..	..	On ships
Bush metal	80	10	5	5	..	..	Bushings
Spring brass	66	33	1	..	..	..	Springs
Britannia metal	2	82	..	..	16	..	Ornaments, etc.
Pewter	2	89	..	..	7	2	Plates, toys, etc.
Muntz metal	60	40	..	..	..	..	Bolts, valves, etc.
Admiralty metal	70	29	1	..	..	..	Ship condensers, etc.

370. Corrosion of Non-ferrous Metals and Their Alloys.—The non-ferrous metals and their alloys offer a very high resistance to corrosion. They may be said to be non-corrodible when compared with iron and steel. Aluminum and nickel are two of the most non-corrosive. Lead will dissolve in water to a slight extent, but this is stopped when the carbonates or sulphates of lime deposited on the lead form a film over the surface. When zinc is exposed to the atmosphere, a thin skin of zinc carbonate soon forms on the surface and prevents further corrosion. Pure tin offers a high resistance to corrosion. Copper will not corrode when exposed to dry air, but it will corrode slightly when there is moisture and other impurities present in the air.

The alloys of the non-ferrous metals offer practically the same resistance to corrosion as do the metals themselves.

CHAPTER XVI

SOME MISCELLANEOUS MATERIALS

371. Paints, Oils, and Varnishes.—A paint is composed of a base, vehicle or binder, and a solvent. Most paints have a stain and a drier added.

The base of a paint is usually white lead or zinc white, though red lead, iron oxide, graphite, etc. are sometimes used. White lead is a combination of lead carbonate and lead hydrate. The former gives the body and the latter the binding properties to the paint. White lead is a good drier of linseed oil, and very little artificial drier is needed. Zinc white is zinc oxide. It is more brilliant but less permanent than white lead. Three coats of white lead are equivalent to five coats of zinc white. Red lead is a double oxide of lead. It makes a good priming paint for new wood and iron. It is anti-corrosive, resists abrasion, and is a good drier of linseed oil. Iron oxides are cheap and serviceable but are not so good as white lead and zinc white. Graphite is opaque and has great covering power.

The usual vehicle or binder is linseed oil. Raw linseed oil is obtained from flaxseed. Boiled linseed oil is made by heating the raw oil alone or with a little drier. Boiled linseed oil dries in half the time required for raw linseed oil, and, consequently, the boiled oil is used more for exterior work while the raw oil is used for interior work. Linseed oil is often adulterated by adding cottonseed, resin, hemp, mineral, or fish oils. Fish oil and cottonseed oil treated with benzine are sometimes used as substitutes for linseed oil.

Spirits of turpentine is the best solvent. It is often adulterated with mineral oils. Benzine and naphtha are sometimes used as substitutes for turpentine.

Staining colors or pigments are used for coloring paints having a white base. The most commonly used pigments are as follows: for black color, soot and charcoal blacks; red color, Venetian red and red lead; brown color, burned umber and raw and burned sierras; green color, chrome green; blue color, Prussian blue or ultramarine blue. By using different amounts and mixtures of the pigments, practically any color and shade may be obtained.

A drier is a lead compound, a manganese compound, or a mixture of these two which is soluble in oil. These compounds act as carriers of oxygen and thus hasten the oxidation and solidification of the linseed oil. Only a very small amount of drier should be used with most paints, and none with "ready mixed" and varnish paints.

Asphalt paints are made by dissolving bitumen in paraffin, petroleum, naphtha, and benzine. These paints are used to protect iron and steel from corrosion.

Varnishes are made by dissolving gums or resins in oil, turpentine, or alcohol. Usually, linseed oil is used with fossil resin, and the compound thinned with turpentine. When the varnish dries, a smooth, solid, and transparent resin is left which forms the varnished surfaces.

The purpose of a paint or varnish is to form an impervious coat over the surface of the material to which it is applied and to keep out the air and moisture and also to give an agreeable color and pleasing appearance to the material painted.

372. Asbestos.—Asbestos is principally a silicate of magnesia with more or less water. It is white in color and fibrous in character. The fibers are hard or soft, depending on the amount of water present; about 11½ per cent or less of water causing hard fibers while about 14 per cent of water makes the fibers soft. Heating asbestos so as to drive out the water makes the fibers very brittle and crumbly.

A chemical analysis of asbestos will give results approximately as follows: silica 40 to 41 per cent; magnesia 41 to 44 per cent; ferrous oxide 1 to 3 per cent; alumina 1 to 3 per cent; and water 11½ to 14½ per cent.

The thermal conductivity of asbestos is about 0.0002.

Asbestos is used for heat insulation purposes, such as covering steam pipes, fire curtains, etc. It comes in the form of fibers, boards, cloth, cylindrical shapes, etc. The fibers may be mixed in a paste and applied to a pipe or other object.

373. Glass.—Ordinary glass is made by fusing together sand or silica with lime, potash, soda, or lead oxide. The process of manufacture is quite complicated.

Glass is a hard, brittle, highly elastic, translucent, and commonly transparent substance with a conchoidal fracture. Glass may be either white or colored. It has a high electrical resistance.

The tensile strength of ordinary glass varies from 2,000 to 3,000 lb. per square inch; the compressive strength from 6,000 to 10,000 lb. per square inch; the cross-bending strength (modulus of rupture) from 3,000 to 4,000 lb. per square inch; and the modulus of elasticity in cross bending from 10,000,000 to 11,000,000 lb. per square inch. The specific gravity of glass varies from 2.4 to 4.5, but is usually between 2.5 and 2.75, giving a weight of from 156 to 172 lb. per cubic foot.

Glass is often classified according to its thickness into "single strength," "double strength," plate glass, etc.; the thickness usually varying from $\frac{1}{16}$ to $\frac{1}{2}$ in.

Glass is used for windows, skylights, insulators, mirrors, bottles, dishes, ornaments, etc.

Wire glass is usually ordinary glass containing a wire mesh in the interior. This wire mesh is usually a fine woven wire which reinforces the glass and prevents a complete failure when the glass is cracked by blows, fire, etc.

374. Glue.—Ordinary glue is an impure gelatine made from bones, skins, or fish and is called bone, skin, or fish glue respectively. A good quality of glue should be free from all specks and grit, have a uniform light-brownish-yellow color, transparent appearance, and glassy fracture. Immersed in cold water, glue will soften and swell without dissolving, and will resume its original properties upon drying. Glue will dissolve in hot water and form a thin syrupy liquid.

The adhesiveness of the different grades of glue varies considerably. The hotter the glue, the better the joint. Remelted glue is not so strong as that freshly prepared, and newly manufactured glue is not so good as old glue. The ultimate adhesive strength of wooden surfaces joined with glue varies from 1,200 to 2,000 lb. per square inch for wood joined end to end across the grain, and from 350 to 1,000 lb. per square inch for wood joined with the grain, depending upon the kind of wood, the quality of the glue, and the workmanship.

Glue is used for joining the surfaces of two materials together as in making wooden joints, etc. Special kinds of glues are manufactured for joining different materials and under different conditions.

375. Rubber.—Almost all of the commercial rubber is derived from a fluid which comes from the outer wood of several species of tropical trees. This fluid (called "latex") is gathered in pails and then coagulated by the application of heat, chemicals, or mechanical manipulation. The coagulated mass is "cured" by smoking, and is then known as "crude" rubber. The crude rubber is washed, shredded, and mixed with sulphur and other ingredients and then rolled into sheets (or pressed into forms) after which it is "vulcanized" by heating under pressure.

The amount of sulphur added to the rubber varies from about 7 to 30 per cent, depending on the hardness desired. Most of the rubber used in engineering contains from 15 to 30 per cent of sulphur. The sulphur also makes the rubber less readily softened by heat or hardened by cold.

The specific gravity of vulcanized rubber is usually taken as an index of its value or quality (though often untrue). The best grades of vulcanized rubber have specific gravities that are less than one.

The stress-strain (load-deformation) curve for a good grade of rubber in tension is very peculiar. Up to about 30 per cent of the ultimate strength there is an increasing rate of elongation with increase of load, then the rate of elongation is constant for a while, and later the rate of elongation decreases with the increase of load. Rubber rarely has an appreciable permanent set, even though stretched nearly to the ultimate before the load is released. The ultimate tensile strength of rubber varies from about 200 to 2,000 lb. per square inch, depending upon the quality.

Good rubber is very flexible and should stretch more than from five to eight times its length without breaking, depending upon the amount of metallic oxides contained in it. The permanent elongation, measured immediately after rupture, of good rubber is usually less than 12 per cent of its original length.

Whenever rubber is subjected to a cycle of stress, where the load is gradually applied and then gradually released, the stress-strain curve exhibits a "mechanical hysteresis" or loss of energy. All of the mechanical energy applied when the rubber is loaded is not returned as such when the load is removed, the energy not returned being dissipated as heat. The faster the rate of loading and unloading, the greater the loss of mechanical energy.

Rubber, unless it is very hard, is so flexible that it has no well defined ultimate strength in compression. The stress-strain curve for good rubber in compression is slightly concave upwards. The unit stress (when the specimen is compressed to about half its original height) varies from about 400 to 1,000 lb. per square inch, depending upon the quality and hardness of the rubber.

Good vulcanized rubber should not harden appreciably when exposed to cold.

The melting point of rubber is about 375 degrees Fahrenheit.

Dilute acids, dilute alkalis, and water have but little effect on rubber.

Oil makes the rubber brittle and shortens its "life."

Vulcanized rubber tends to become hard and brittle with the lapse of time and to lose its "life" or ability to stretch. The loss of "life" or the ability to stretch is hastened by the presence of metallic oxides (such as chalk or zinc oxide) in the vulcanized rubber.

Rubber is used for belting, electrical insulation, floor coverings, vehicle tires (solid and pneumatic), toilet articles, rulers and scales, articles of clothing, rubber bands, erasers, hose, and many other purposes.

376. Leather.—Leather is an imputrescible material made from the hides and skins of living creatures by chemical and mechanical treatment.

Leather may be classified according to the process of manufacture as follows:

1. Tanned leather in which the hides and skins are combined with tannin or tannic acid.
2. Tanned leather in which the skins are prepared with mineral salts.
3. Chamoised leather in which the skins are rendered imputrescible by treatment with oils or fats, the decomposition products of which are the actual tanning agents.

Leather may also be classified according to the kind of animal (ox, heifer, cow, bull, calf, goat, sheep, horse, etc.) it comes from and also according to the country or part of the world where the animals were raised. Ox and heifer hides are the best, followed by cow and bull hides respectively.

The properties of leather are affected by the processes of tanning, finishing, currying (working oil or grease into the leather), etc., in preparing the hides for use. The average tensile strength of leather is from 2,000 to 3,000 lb. per square inch. The weight of well-tanned leather is about 62½ lb. per cubic foot.

Leather is used for shoes, belting, gloves, clothes, book binding, bags, trunks, etc.

Rawhide is untanned leather. When sound, rawhide is about one and a half times as strong as tanned leather and offers a greater resistance to impact loads. Rawhide is used for textile machinery connections, looms, ships, tiller ropes, etc.

377. Paper.—Paper may be classified according to use into four classes, namely: (1) writing and drawing papers, (2) printing and newspapers, (3) wrapping papers, and (4) tissue and cigarette papers.

The process of paper manufacture consists essentially of two main divisions: (1) The treatment of the raw material, including cleaning, dusting, boiling, washing, bleaching, and reducing to a pulp; (2) the methods by which the prepared pulp or fibers are converted into paper ready for the market, and these include the operations of beating, sizing, coloring, surfacing, cutting, etc.

The materials most commonly used in the manufacture of paper are rags, straw, and wood.

Most of the paper used for structural purposes comes under the third class. Ordinary building paper is a wrapping paper which is placed in the walls or roofs as an added protection against the passage of air and heat. Tarred paper is paper covered or soaked with tar which tends to prevent the passage of moisture. Sometimes an ornamented or decorated paper is pasted on the walls and ceilings of rooms to improve their appearance.

378. Canvas.—Canvas is a heavy, strong, coarse cloth made from flax, hemp, tow, jute, or cotton. The cloth may be of a natural color, bleached, or dyed. Common colors are white and tan.

Two or more ply yarns are used, and the cloth is woven loosely or closely as desired. The quality is determined by the yarn used, the closeness of weaving, and the weight and strength in tension. The tensile strength of canvas varies from 3,000 to 10,000 lb. per square inch.

Canvas is used for bags, tarpaulins, sails, belts, tents, etc.

379. Ropes.—Ropes are said to include all large cords more than a half inch in diameter.

Ordinary rope is made by spinning the fibers of the material together to make a yarn, twisting from twenty to eighty pieces of yarn together to form a strand, and then twisting three or four strands to form a 3 or 4 ply rope. In each case of twisting, the direction of the twist is reversed so that the rope will not untwist when loaded.

Wire rope is made by twisting 4, 7, 12, 19, or 37 wires together to form a strand, and then twisting a number of strands together to form a rope. Frequently, a hemp or manila rope is used for a core. Hoisting ropes usually have 19 wires to the strand, while 7 or 12 wires to the strand are used for standing ropes, rigging, etc.

Great care must be taken in tying or splicing ropes together so that the strength will not be reduced too much. A good splice or knot should have about 80 per cent of the strength of the rope.

The factor of safety for rope varies from about 5 to 35, depending upon the kinds of loads, the speed at which the ropes are worked, the diameter of the pulley blocks, etc.

The following table gives approximate values of the ultimate tensile strength of several kinds of rope. The unit stresses are based on the original gross cross-sectional area of the rope.

TENSILE STRENGTH OF ROPE

KIND OF ROPE	TENSILE STRENGTH IN POUNDS PER SQUARE INCH
Wire (iron and steel)	50,000 to 90,000
Cotton	5,000 to 8,000
Hemp	5,000 to 8,000
Manila	5,000 to 10,000

380. Belts.—Rubber belts are made by weaving cotton canvas of the required length and width and then covering the canvas with vulcanized rubber. The belts may be of 2, 3, or 4 ply (thicknesses or layers of canvas and rubber) according to the power to be transmitted. Rubber belts are stronger than leather belts, usually run smoother and truer, have a higher coefficient of friction, and are impervious to water, but they may be injured more by slipping when overloaded. The tensile strength of rubber belting varies from about 2,200 to 3,800 lb. per square inch on the average.

Leather belts should be made from the best quality of well tanned ox hides. The hides should be cut in strips from 4 to 6 ft. long and about 3/16 in. thick, and then scarfed, spliced, and cemented end to end to make the desired length of belt. According to the strength required, the strips of leather are riveted or cemented together in thicknesses to form "single" or "double"

18

strength belts. For light loads, the "single" strength belt gives the greater adhesion, but the "double" strength belt is necessary for heavy loads. The "flesh" side of the belt is usually placed next to the pulley as it gives the best wear, though the "grain" side of the belt gives the best adhesion. Tensile tests on leather belts have given results varying from 1,500 to 5,500 lb. per square inch, but the strength of the average belt is probably nearer the lower value.

Canvas belts are sometimes used, but they give much poorer service than good rubber or leather belts as the adhesion and life are less. The strength of canvas belts in tension varies from 3,000 to 8,000 lb. per square inch.

Great care must be taken in splicing belts so as to secure a good joint. Most canvas, rubber, and leather belts are spliced by lacing or riveting and rarely by sewing or gluing. Splicing a belt reduces it's strength about one-half; consequently, a belt should be made of the desired length so as to require no shortening or lengthening.

Thin flat bands of tempered steel have been used for belts in special cases. These belts are very strong, have but little slip, have a very high efficiency, may be run at very high speeds, and require about 1/5 of the width of a good leather belt for the same power.

APPENDIX A

AMERICAN SOCIETY FOR TESTING MATERIALS

PHILADELPHIA, PA., U. S. A.

AFFILIATED WITH THE

INTERNATIONAL ASSOCIATION FOR TESTING MATERIALS

STANDARD SPECIFICATIONS AND TESTS FOR PORTLAND CEMENT

Serial Designation: C 9–21

These specifications and tests are issued under the fixed designation C 9; the final number indicates the year of original adoption as standard, or in the case of revision, the year of last revision.

ADOPTED, 1904; REVISED, 1908, 1909, 1916, 1920 (EFFECTIVE JAN. 1, 1921)

These specifications were approved January 15, 1921, as "Tentative Standard" by the American Engineering Standards Committee

---

SPECIFICATIONS

1. Definition.—Portland cement is the product obtained by finely pulverizing clinker produced by calcining to incipient fusion an intimate and properly proportioned mixture of argillaceous and calcareous materials, with no additions subsequent to calcination excepting water and calcined or uncalcined gypsum.

I. CHEMICAL PROPERTIES

2. Chemical Limits.—The following limits shall not be exceeded:

Loss on ignition, per cent	4.00
Insoluble residue, per cent	0.85
Sulfuric anhydride (SO_3), per cent	2.00
Magnesia (MgO), per cent	5.00

II. PHYSICAL PROPERTIES

3. Specific Gravity.—The specific gravity of cement shall be not less than 3.10 (3.07 for white portland cement). Should the test of cement as received fall below this requirement a second test may be made upon an ignited sample. The specific gravity test will not be made unless specifically ordered.

4. Fineness.—The residue on a standard No. 200 sieve shall not exceed 22 per cent by weight.

5. Soundness.—A pat of neat cement shall remain firm and hard, and show no signs of distortion, cracking, checking, or disintegration in the steam test for soundness.

6. Time of Setting.—The cement shall not develop initial set in less than 45 minutes when the Vicat needle is used or 60 minutes when the Gillmore needle is used. Final set shall be attained within 10 hours.

7. Tensile Strength.—The average tensile strength in pounds per square inch of not less than three standard mortar briquettes (see Section 50) composed of 1 part cement and 3 parts standard sand, by weight, shall be equal to or higher than the following:

Age at test, days	Storage of briquettes	Tensile strength, pounds per square inch
7	1 day in moist air, 6 days in water...........	200
28	1 day in moist air, 27 days in water...........	300

8. The average tensile strength of standard mortar at 28 days shall be higher than the strength at 7 days.

III. Packages, Marking and Storage

9. Packages and Marking.—The cement shall be delivered in suitable bags or barrels with the brand and name of the manufacturer plainly marked thereon, unless shipped in bulk. A bag shall contain 94 lb. net. A barrel shall contain 376 lb. net.

10. Storage.—The cement shall be stored in such a manner as to permit easy access for proper inspection and identification of each shipment, and in a suitable weather-tight building which will protect the cement from dampness.

IV. Inspection

11. Inspection.—Every facility shall be provided the purchaser for careful sampling and inspection at either the mill or at the site of the work, as may be specified by the purchaser. At least 10 days from the time of sampling shall be allowed for the completion of the 7-day test, and at least 31 days shall be allowed for the completion of the 28-day test. The cement shall be tested in accordance with the methods hereinafter prescribed. The 28-day test shall be waived only when specifically so ordered.

V. Rejection

12. Rejection.—The cement may be rejected if it fails to meet any of the requirements of these specifications.

13. Cement shall not be rejected on account of failure to meet the fineness requirement if upon retest after drying at 100 degrees Centigrade for 1 hour it meets this requirement.

14. Cement failing to meet the test for soundness in steam may be accepted if it passes a retest using a new sample at any time within 28 days thereafter.

15. Packages varying more than 5 per cent from the specified weight may be rejected; and if the average weight of packages in any shipment, as shown by weighing 50 packages taken at random, is less than that specified, the entire shipment may be rejected.

TESTS

VI. Sampling

16. Number of Samples.—Tests may be made on individual or composite samples as may be ordered. Each test sample should weigh at least 8 lb.

17. (*a*) *Individual Sample.*—If sampled in cars, one test sample shall be taken from each 50 bbl. or fraction thereof. If sampled in bins, one sample shall be taken from each 100 bbl.

(*b*) *Composite Sample.*—If sampled in cars, one sample shall be taken from one sack in each 40 sacks (or 1 bbl. in each 10 bbl.) and combined to form one test sample. If sampled in bins or warehouses, one test sample shall represent not more than 200 bbl.

18. Method of Sampling.—Cement may be sampled at the mill by any of the following methods that may be practicable, as ordered:

(*a*) *From the Conveyor Delivering to the Bin.*—At least 8 lb. of cement shall be taken from approximately each 100 bbl. passing over the conveyor.

(*b*) *From Filled Bins by Means of Proper Sampling Tubes.*—Tubes inserted vertically may be used for sampling cement to a maximum depth of 10 ft. Tubes inserted horizontally may be used where the construction of the bin permits. Samples shall be taken from points well distributed over the face of the bin.

(*c*) *From Filled Bins at Points of Discharge.*—Sufficient cement shall be drawn from the discharge openings to obtain samples representative of the cement contained in the bin, as determined by the appearance at the discharge openings of indicators placed on the surface of the cement directly above these openings before drawing of the cement is started.

19. Treatment of Sample.—Samples preferably shall be shipped and stored in air-tight containers. Samples shall be passed through a sieve having 20 meshes per linear inch in order to thoroughly mix the sample, break up lumps, and remove foreign materials.

VII. Chemical Analysis

Loss on Ignition

20. Method.—One gram of cement shall be heated in a weighed covered platinum crucible, of 20 to 25 c.c. capacity, as follows, using either method (*a*) or (*b*) as ordered:

(*a*) The crucible shall be placed in a hole in an asbestos board, clamped horizontally so that about three-fifths of the crucible projects below, and blasted at a full red heat for 15 minutes with an inclined flame; the loss in weight shall be checked by a second blasting for 5 minutes. Care shall be taken to wipe off particles of asbestos that may adhere to the crucible when withdrawn from the hole in the board. Greater neatness and shortening of the time of heating are secured by making a hole to fit the crucible in a circular disk of sheet platinum and placing this disk over a somewhat larger hole in an asbestos board.

(*b*) The crucible shall be placed in a muffle at any temperature between 900 and 1,000 degrees Centigrade for 15 minutes and the loss in weight shall be checked by a second heating for 5 minutes.

21. Permissible Variation.—A permissible variation of 0.25 will be allowed, and all results in excess of the specified limit but within this permissible variation shall be reported as 4 per cent.

Insoluble Residue

22. Method.—To a 1-gram sample of cement shall be added 10 c.c. of water and 5 c.c. of concentrated hydrochloric acid; the liquid shall be warmed until effervescence ceases. The solution shall be diluted to 50 c.c. and digested on a steam bath or hot plate until it is evident that decomposition of the cement is complete. The residue shall be filtered, washed with cold water, and the filter paper and contents digested in about 30 c.c. of a 5 per cent solution of sodium carbonate, the liquid being held at a temperature just short of boiling for 15 minutes. The remaining residue shall be filtered, washed with cold water, then with a few drops of hot hydrochloric acid, 1:9, and finally with hot water, and then ignited at a red heat and weighed as the insoluble residue.

23. Permissible Variation.—A permissible variation of 0.15 will be allowed, and all results in excess of the specified limit but within this permissible variation shall be reported as 0.85 per cent.

Sulfuric Anhydride

24. Method.—One gram of the cement shall be dissolved in 5 c.c. of concentrated hydrochloric acid diluted with 5 c.c. of water, with gentle warming; when solution is complete, 40 c.c. of water shall be added, the solution filtered, and the residue washed thoroughly with water. The solution shall be diluted to 250 c.c., heated to boiling, and 10 c.c. of a hot 10-per-cent solution of barium chloride shall be added slowly, drop by drop, from a pipette and the boiling continued until the precipitate is well formed. The solution shall be digested on the steam bath until the precipitate has settled. The precipitate shall be filtered, washed, and the paper and contents placed in a weighed platinum crucible and the paper slowly charred and consumed without flaming. The barium sulfate shall then be ignited and weighed. The weight obtained multiplied by 34.3 gives the percentage of sulfuric anhydride. The acid filtrate obtained in the determination of the insoluble residue may be used for the estimation of sulfuric anhydride instead of using a separate sample.

25. Permissible Variation.—A permissible variation of 0.10 will be allowed, and all results in excess of the specified limit but within this permissible variation shall be reported as 2.00 per cent.

Magnesia

26. Method.—To 0.5 gram of the cement in an evaporating dish shall be added 10 c.c. of water to prevent lumping and then 10 c.c. of concentrated hydrochloric acid. The liquid shall be gently heated and agitated until attack is complete. The solution shall then be evaporated to complete dryness on a steam or water bath. To hasten dehydration the residue may be heated to 150 or even 200 degrees Centigrade for ½ to 1 hour. The residue shall be treated with 10 c.c. of concentrated hydrochloric acid diluted with an equal amount of water. The dish shall be covered and the solution digested for 10 minutes on a steam bath or water bath. The diluted solution shall be filtered and the separated silica washed thoroughly with water.[1] Five cubic centimeters of concentrated hydrochloric acid and sufficient bromine water to precipitate any manganese which may be present shall be added to the filtrate (about 250 c.c.). This shall be made alkaline with ammonium hydroxide, boiled until there is but a faint odor of ammonia, and the precipitated iron and aluminum hydroxides, after settling, shall be washed with hot water, once by decantation and slightly on the filter. Setting aside the filtrate, the precipitate shall be transferred by a jet of hot water to the precipitating vessel and dissolved in 10 c.c. of hot hydrochloric acid. The paper shall be extracted with acid, the solution and washings being added to the main solution. The aluminum and iron shall then be reprecipitated at boiling heat by ammonium hydroxide and bromine water in a volume of about 100 c.c., and the second precipitate shall be collected and washed on the filter used in the first instance if this is still intact. To the combined filtrates from the hydroxides of iron and aluminum, reduced in volume if need be, 1 c.c. of ammonium hydroxide shall be added, the solution brought to boiling, 25 c.c. of a saturated solution of boiling ammonium oxalate added, and the boiling continued until the precipitated calcium oxalate has assumed a well-defined granular form. The precipitate after one hour shall be filtered and washed, then with the filter shall be placed wet in a platinum crucible, and the paper burned off over a small flame of a bunsen burner; after ignition it shall be redissolved in hydrochloric acid and the solution diluted to 100 c.c. Ammonia shall be added in slight excess, and the liquid boiled. The lime shall then be reprecipitated by ammonium oxalate, allowed to stand until settled, filtered, and washed. The combined filtrates from the calcium precipitates shall be acidified with hydrochloric acid, concentrated on the steam bath to about 150 c.c., and made slightly alkaline with ammonium hydroxide, boiled and filtered (to remove a little aluminum and iron and perhaps calcium). When cool, 10 c.c. of saturated solution of sodium-ammonium-hydrogen phosphate shall be added with constant stirring. When the crystalline ammonium-magnesium orthophosphate has formed, ammonia shall be added in moderate excess. The solution shall be set aside for several hours in a cool place,

[1] Since this procedure does not involve the determination of silica, a second evaporation is unnecessary.

filtered, and washed with water containing 2.5 per cent of NH_3. The precipitate shall be dissolved in a small quantity of hot hydrochloric acid, the solution diluted to about 100 c.c., 1 c.c. of a saturated solution of sodium-

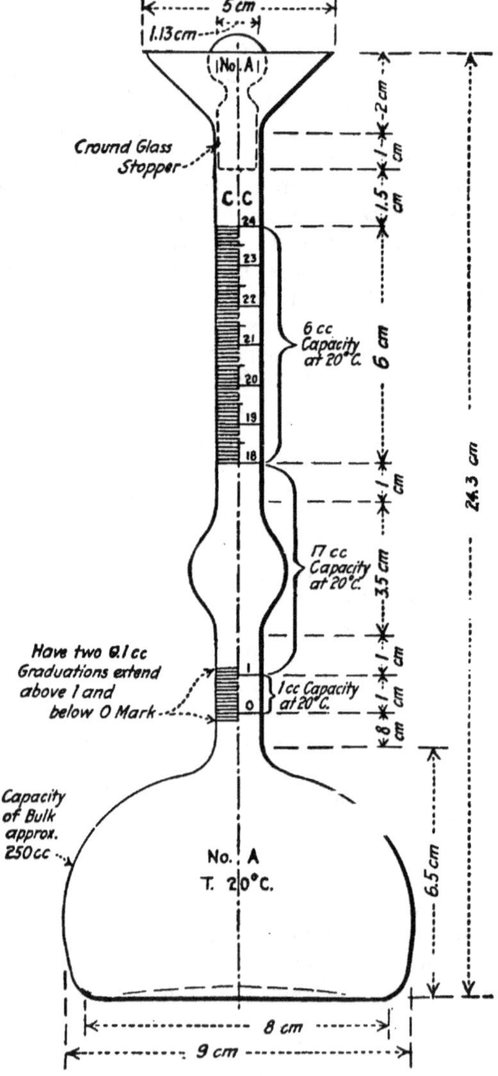

FIG. 1.—Le Chatelier apparatus.

ammonium-hydrogen phosphate added, and ammonia drop by drop, with constant stirring, until the precipitate is again formed as described and the ammonia is in moderate excess. The precipitate shall then be allowed to stand about two hours, filtered, and washed as before. The paper and contents shall be placed in a weighed platinum crucible, the paper slowly charred, and the resulting carbon carefully burned off. The precipitate shall then be ignited to constant weight over a Meker burner, or a blast not strong enough to soften or melt the pyrophosphate. The weight of magnesium pyrophosphate obtained multiplied by 72.5 gives the percentage of magnesia. The precipitate so obtained always contains some calcium and usually small quantities of iron, aluminum, and manganese as phosphates.

27. Permissible Variation.—A permissible variation of 0.4 will be allowed, and all results in excess of the specified limit but within this permissible variation shall be reported as 5.00 per cent.

VIII. Determination of Specific Gravity

28. Apparatus.—The determination of specific gravity shall be made with a standardized *Le* Chatelier apparatus which conforms to the requirements illustrated in Fig. 1. This apparatus is standardized by the United States Bureau of Standards. Kerosene free from water, or benzine not lighter than 62 degrees Baumé, shall be used in making this determination.

29. Method.—The flask shall be filled with either of these liquids to a point on the stem between zero and 1 c.c.; and 64 grams of cement, of the same temperature as the liquid, shall be slowly introduced, taking care that the cement does not adhere to the inside of the flask above the liquid and to free the cement from air by rolling the flask in an inclined position. After all the cement is introduced, the level of the liquid will rise to some division of the graduated neck; the difference between readings is the volume displaced by 64 grams of the cement.

The specific gravity shall then be obtained from the formula

$$\text{Specific gravity} = \frac{\text{Weight of cement (grams)}}{\text{Displaced volume (cubic centimeters)}}$$

30. The flask, during the operation, shall be kept immersed in water, in order to avoid variations in the temperature of the liquid in the flask, which shall not exceed 0.5 degree Centigrade. The results of repeated tests should agree within 0.01.

31. The determination of specific gravity shall be made on the cement as received; if it falls below 3.10, a second determination shall be made after igniting the sample as described in Section 20.

IX. Determination of Fineness

32. Apparatus.—Wire cloth for standard sieves for cement shall be woven (not twilled) from brass, bronze, or other suitable wire, and mounted without distortion on frames not less than 1½ in. below the top of the frame. The sieve frames shall be circular, approximately 8 in. in diameter, and may be provided with a pan and cover.

33. A standard No. 200 sieve is one having nominally an 0.0029-in. opening and 200 wires per inch standardized by the United States Bureau of Standards, and conforming to the following requirements

The No. 200 sieve should have 200 wires per inch, and the number of wires in any whole inch shall not be outside the limits of 192 to 208. No opening between adjacent parallel wires shall be more than 0.0050 in. in width. The diameter of the wire should be 0.0021 in. and the average diameter shall not be outside the limits 0.0019 to 0.0023 in. The value of the sieve as determined by sieving tests made in conformity with the standard specification for these tests on a standardized cement which gives a residue of 25 to 20 per cent on the No. 200 sieve, or on other similarly graded material, shall not show a variation of more than 1.5 per cent above or below the standards maintained at the Bureau of Standards.

34. Method.—The test shall be made with 50 grams of cement. The sieve shall be thoroughly clean and dry. The cement shall be placed on the No. 200 sieve, with pan and cover attached, if desired, and shall be held in one hand in a slightly inclined position so that the sample will be well distributed over the sieve, at the same time gently striking the side about 150 times per minute against the palm of the other hand on the upstroke. The sieve shall be turned every 25 strokes about one-sixth of a revolution in the same direction. The operation shall continue until not more than 0.05 gram passes through in 1 minute of continuous sieving. The fineness shall be determined from the weight of the residue on the sieve expressed as a percentage of the weight of the original sample.

35. Mechanical sieving devices may be used, but the cement shall not be rejected if it meets the fineness requirement when tested by the hand method described in Section 34.

X. Mixing Cement Pastes and Mortars

36. Method.—The quantity of dry material to be mixed at one time shall not exceed 1,000 grams nor be less than 500 grams. The proportions of cement or cement and sand shall be stated by weight in grams of the dry materials; the quantity of water shall be expressed in cubic centimeters (1 c.c. of water = 1 gram). The dry materials shall be weighed, placed upon a non-absorbent surface, thoroughly mixed dry if sand is used, and a crater formed in the center, into which the proper percentage of clean water shall be poured; the material on the outer edge shall be turned into the crater by the aid of a trowel. After an interval of ½ minute for the absorption of the water the operation shall be completed by continuous, vigorous mixing, squeezing, and kneading with the hands for at least 1 minute.[1] During the operation of mixing, the hands should be protected by rubber gloves.

37. The temperature of the room and the mixing water shall be maintained as nearly as practicable at 21 degrees Centigrade (70 degrees Fahrenheit).

[1] In order to secure uniformity in the results of tests for the time of setting and tensile strength the manner of mixing above described should be carefully followed. At least one minute is necessary to obtain the desired plasticity which is not appreciably affected by continuing the mixing for several minutes. The exact time necessary is dependent upon the personal equation of the operator. The error in mixing should be on the side of overmixing.

XI. NORMAL CONSISTENCY

38. Apparatus.—The Vicat apparatus consists of a frame *A* (Fig. 2) bearing a movable rod *B*, weighing 300 grams, one end *C* being 1 cm. in diameter for a distance of 6 cm., the other having a removable needle D, 1 mm. in diameter, 6 cm. long. The rod is reversible, and can be held in any desired position by a screw *E*, and has midway between the ends a mark *F* which

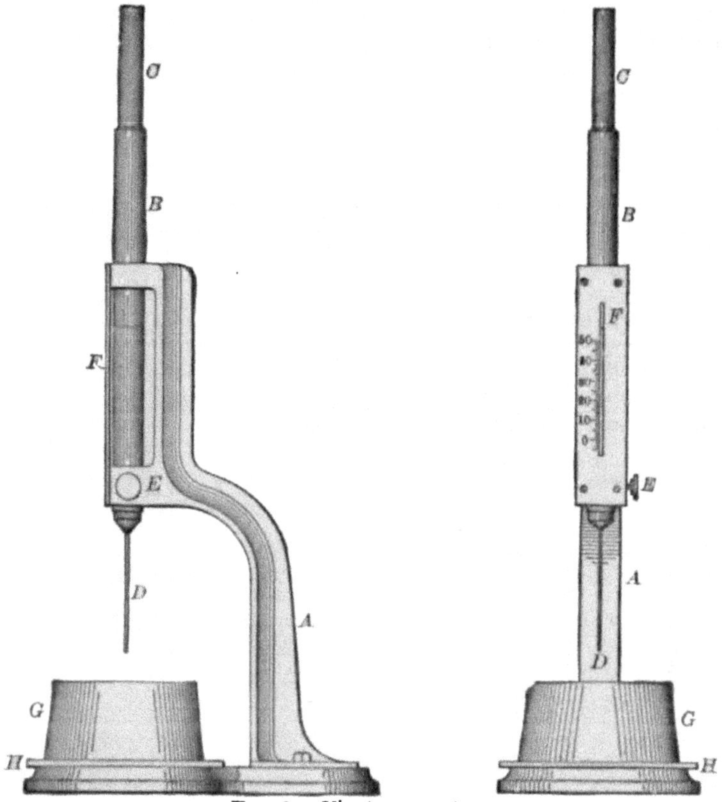

FIG. 2.—Vicat apparatus.

moves under a scale (graduated to millimeters) attached to the frame *A*. The paste is held in a conical, hard-rubber ring *G*, 7 cm. in diameter at the base, 4 cm. high, resting on a glass plate *H* about 10 cm. square.

39. Method.—In making the determination, 500 grams of cement, with a measured quantity of water, shall be kneaded into a paste, as described in Section 36, and quickly formed into a ball with the hands, completing the operation by tossing it six times from one hand to the other, maintained about 6 in. apart; the ball resting in the palm of one hand shall be pressed into the larger end of the rubber ring held in the other hand, completely filling the ring with paste; the excess at the larger end shall then be removed by a single movement of the palm of the hand; the ring shall then be placed on its larger end on a glass plate and the excess paste at the smaller end sliced off at the top of the ring by a single oblique stroke of a trowel held at a slight angle with the top of the ring. During these operations care shall be taken not to compress the paste. The paste confined in the ring, resting on the plate, shall be placed under the rod, the larger end of which shall be brought in contact with the surface of the paste; the scale shall be then read, and the rod quickly released. The paste shall be of normal consistency when the rod settles to a point 10 mm. below the original surface in ½ minute after being released. The apparatus shall be free from all vibrations during the test. Trial pastes shall be made with varying percentages of water until the normal consistency is obtained. The amount of water required shall be expressed in percentage by weight of the dry cement.

40. The consistency of standard mortar shall depend on the amount of water required to produce a paste of normal consistency from the same sample of cement. Having determined the normal consistency of the sample, the consistency of standard mortar made from the same sample shall be as indicated in Table I, the values being in percentage of the combined dry weights of the cement and standard sand.

TABLE I.—PERCENTAGE OF WATER FOR STANDARD MORTARS

Percentage of water for neat cement paste of normal consistency	Percentage of water for one cement, three standard Ottawa sand	Percentage of water for neat cement paste of normal consistency	Percentage of water for one cement, three standard Ottawa sand
15	9.0	23	10.3
16	9.2	24	10.5
17	9.3	25	10.7
18	9.5	26	10.8
19	9.7	27	11.0
20	9.8	28	11.2
21	10.0	29	11.3
22	10.2	30	11.5

XII. DETERMINATION OF SOUNDNESS[1]

41. Apparatus.—A steam apparatus, which can be maintained at a temperature between 98 and 100 degrees Centigrade, or one similar to that shown in Fig. 3, is recommended. The capacity of this apparatus may be increased by using a rack for holding the pats in a vertical or inclined position.

[1] Unsoundness is usually manifested by change in volume which causes distortion, cracking, checking, or disintegration.

Pats improperly made or exposed to drying may develop what are known as shrinkage cracks within the first 24 hours and are not an indication of unsoundness. These conditions are illustrated in Fig. 4.

The failure of the pats to remain on the glass or the cracking of the glass to which the pats are attached does not necessarily indicate unsoundness.

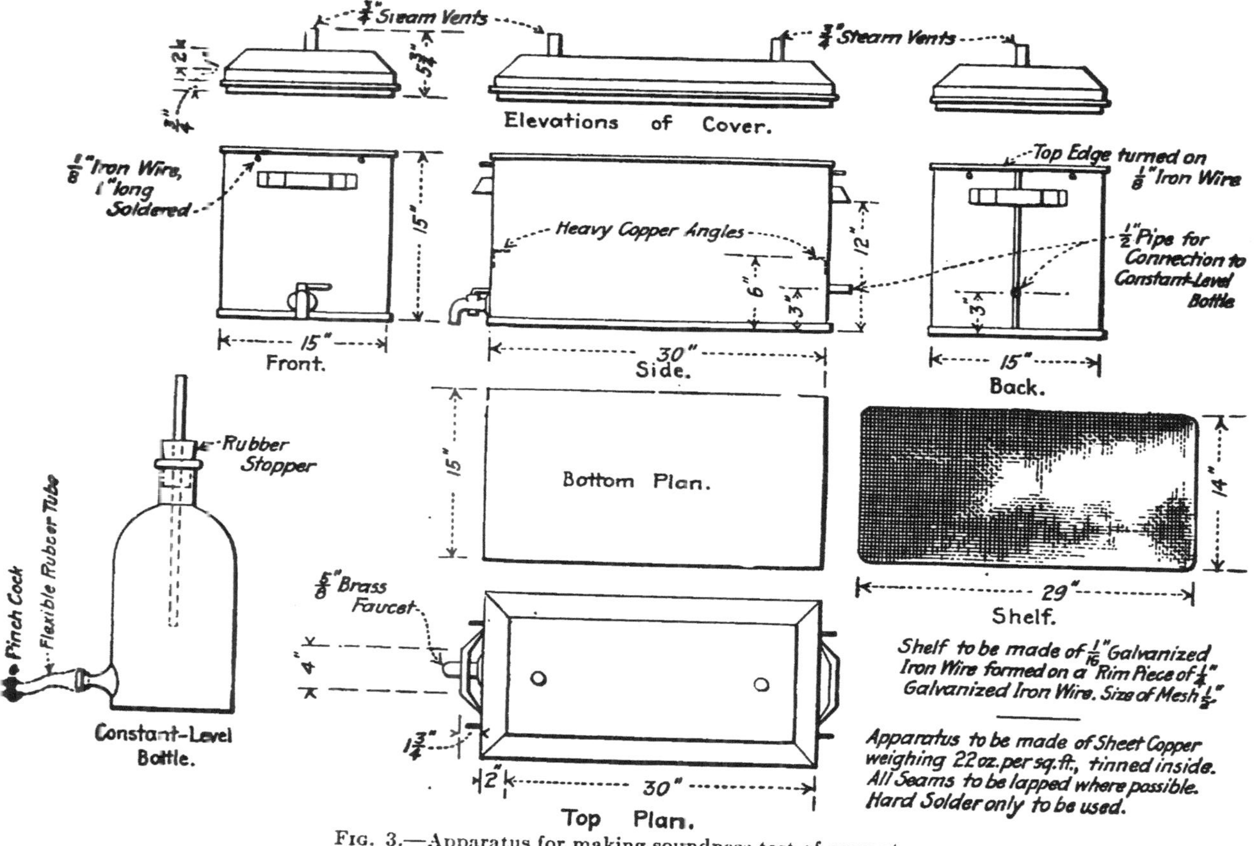

FIG. 3.—Apparatus for making soundness test of cement.

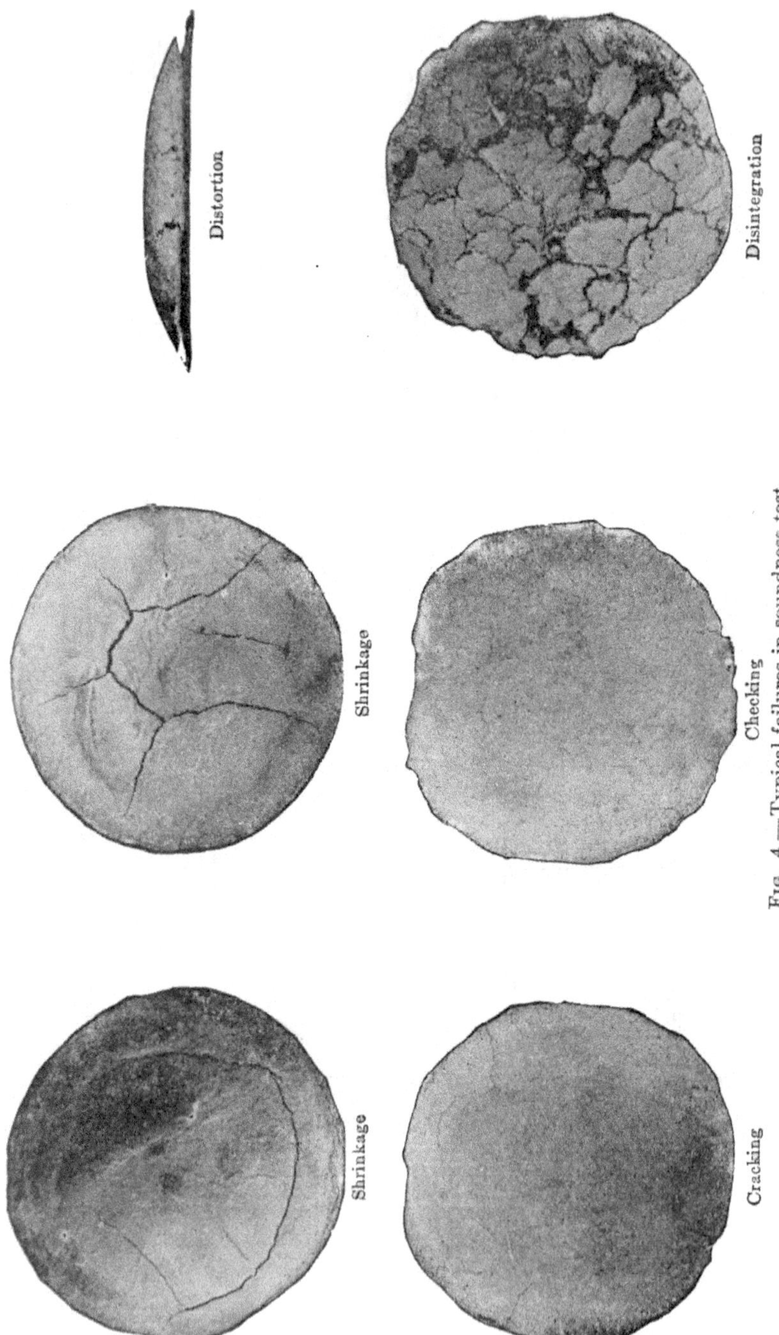

FIG. 4.—Typical failures in soundness test.

42. Method.—A pat from cement paste of normal consistency about 3 in. in diameter, ½ in. thick at the center, and tapering to a thin edge, shall be made on clean glass plates about 4 in. square, and stored in moist air for 24 hours. In molding the pat, the cement paste shall first be flattened on the glass and the pat then formed by drawing the trowel from the outer edge toward the center.

43. The pat shall then be placed in an atmosphere of steam at a temperature between 98 and 100 degrees Centigrade upon a suitable support 1 in. above boiling water for 5 hours.

44. Should the pat leave the plate, distortion may be detected best with a straight edge applied to the surface which was in contact with the plate.

XIII. Determination of Time of Setting

45. The following are alternate methods, either of which may be used as ordered:

(a) Pat with top surface flattened for determining time of setting by Gillmore method.

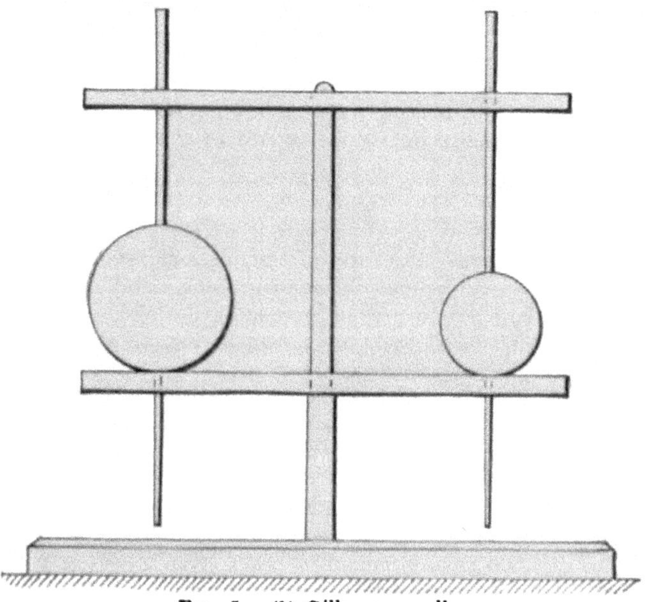

Fig. 5.—(b) Gillmore needles.

46. Vicat Apparatus.—The time of setting shall be determined with the Vicat apparatus described in Section 38 (see Fig. 2).

47. Vicat Method.—A paste of normal consistency shall be molded in the hard-rubber ring *G* as described in Section 39, and placed under the rod *B*, the smaller end of which shall then be carefully brought in contact with the surface of the paste, and the rod quickly released. The initial set shall be said to have occurred when the needle ceases to pass a point 5 mm. above the glass plate in ½ minute after being released; and the final set, when the needle does not sink visibly into the paste. The test pieces shall be kept in moist air during the test. This may be accomplished by placing them on a rack over water contained in a pan and covered by a damp cloth, kept from contact with them by means of a wire screen; or they may be stored in a moist closet. Care shall be taken to keep the needle clean, as the collection of cement on the sides of the needle retards the penetration, while cement on the point may increase the penetration. The time of setting is affected not only by the percentage and temperature of the water used and the amount of kneading the paste receives, but by the temperature and humidity of the air; and its determination is, therefore, only approximate.

48. Gillmore Needles.—The time of setting shall be determined by the Gillmore needles. The Gillmore needles should preferably be mounted as shown in Fig. 5 (*b*).

49. Gillmore Method.—The time of setting shall be determined as follows: A pat of neat cement paste about 3 in. in diameter and ½ in. in thickness with a flat top (Fig. 5 (*a*)), mixed to a normal consistency, shall be kept in moist air at a temperature maintained as nearly as practicable at 21 degrees Centigrade (70 degrees Fahrenheit). The cement shall be considered to have acquired its initial set when the pat will bear, without appreciable indentation, the Gillmore needle 1/12 in. in diameter, loaded to weigh ¼ lb. The final set has been acquired when the pat will bear, without appreciable indentation, the Gillmore needle 1/24 in. in diameter, loaded to weigh 1 lb. In making the test, the needles shall be held in a vertical position and applied lightly to the surface of the pat.

XIV. Tension Tests

50. Form of Test Piece.—The form of test piece shown in Fig. 6 shall be used. The molds shall be made of non-corroding metal and have sufficient material in the sides to prevent spreading during molding. Gang molds when used shall be of the type shown in Fig. 7. Molds shall be wiped with an oily cloth before using.

51. Standard Sand.—The sand to be used shall be natural sand from Ottawa, Ill., screened to pass a No. 20 sieve and retained on a No. 30 sieve. This sand may be obtained from the Ottawa Silica Co., at a cost of 3 cents per pound, f.o.b. cars, Ottawa, Ill.

52. This sand, having passed the No. 20 sieve, shall be considered standard when not more than 5 grams pass the No. 30 sieve after 1 minute continuous sieving of a 500-gram sample.

53. The sieves shall conform to the following specifications:

The No. 20 sieve shall have between 19.5 and 20.5 wires per whole inch of the warp wires and between 19 and 21 wires per whole inch of the shoot wires. The diameter of the wire should be 0.0165 in., and the average diameter shall not be outside the limits of 0.0160 and 0.0170 in.

The No. 30 sieve shall have between 29.5 and 30.5 wires per whole inch of the warp wires and between 28.5 and 31.5 wires per whole inch of the shoot wires. The diameter of the wire should be 0.0110 in., and the average diameter shall not be outside the limits 0.0105 to 0.0115 in.

54. Molding.—*I*mmediately after mixing, the standard mortar shall be placed in the molds, pressed in firmly with the thumbs, and smoothed off with a trowel without ramming. Additional mortar shall be heaped above the mold and smoothed off with a trowel; the trowel shall be drawn over

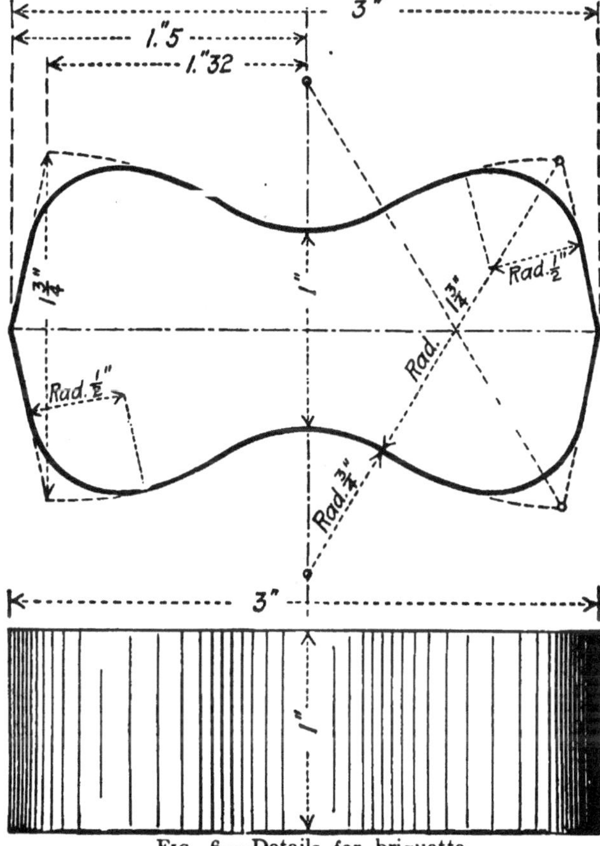

Fig. 6.—Details for briquette.

the mold in such a manner as to exert a moderate pressure on the material. The mold shall then be turned over and the operation of heaping, thumbing, and smoothing off repeated.

55. Testing.—Tests shall be made with any standard machine. The briquettes shall be tested as soon as they are removed from the water. The bearing surfaces of the clips and briquettes shall be free from grains of sand or dirt. The briquettes shall be carefully centered and the load applied continuously at the rate of 600 lb. per minute.

56. Testing machines should be frequently calibrated in order to determine their accuracy.

19

57. Faulty Briquettes.—Briquettes that are manifestly faulty, or which give strengths differing more than 15 per cent from the average value of all

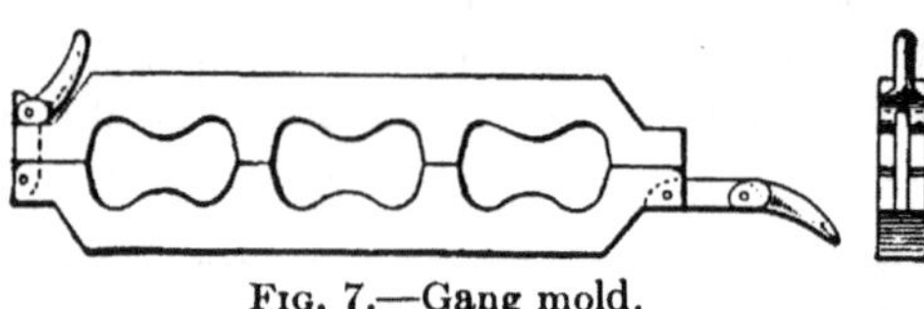

Fig. 7.—Gang mold.

test pieces made from the same sample and broken at the same period, shall not be considered in determining the tensile strength.

XV. Storage of Test Pieces

58. Apparatus.—The moist closet may consist of a soapstone, slate, or concrete box, or a wooden box lined with metal. If a wooden box is used, the interior should be covered with felt or broad wicking kept wet. The bottom of the moist closet should be covered with water. The interior of the closet should be provided with non-absorbent shelves on which to place the test pieces, the shelves being so arranged that they may be withdrawn readily.

59. Methods.—Unless otherwise specified all test pieces, immediately after molding, shall be placed in the moist closet for from 20 to 24 hours.

60. The briquettes shall be kept in molds on glass plates in the moist closet for at least 20 hours. After 24 hours in moist air the briquettes shall be immersed in clean water in storage tanks of non-corroding material.

61. The air and water shall be maintained as nearly as practicable at a temperature of 21 degrees Centigrade (70 degrees Fahrenheit).

APPENDIX B

AMERICAN SOCIETY FOR TESTING MATERIALS

PHILADELPHIA, PA., U. S. A.

AFFILIATED WITH THE

INTERNATIONAL ASSOCIATION FOR TESTING MATERIALS

STANDARD SPECIFICATIONS
FOR
STRUCTURAL STEEL FOR BUILDINGS

Serial Designation: A 9–16

These specifications are issued under the fixed designation A 9; the final number indicates the year of original adoption as standard, or in the case of revision, the year of last revision.

ADOPTED, 1901; REVISED, 1909, 1913, 1914, 1916

In view of the abnormal difficulty in obtaining materials in time of war, the rejection limits for sulphur in all steels and for phosphorus in acid steels shall be raised 0.01 per cent above the values given in these specifications. This shall be effective during the period of the war and until otherwise ordered by the Society.

I. MANUFACTURE

1. Process.—(*a*) Structural steel, except as noted in Paragraph (*b*), may be made by the bessemer or the open-hearth process.

(*b*) Rivet steel, and steel for plates or angles over ¾ in. in thickness which are to be punched, shall be made by the open-hearth process.

II. CHEMICAL PROPERTIES AND TESTS

2. Chemical Composition.—The steel shall conform to the following requirements as to chemical composition:

		STRUCTURAL STEEL, PER CENT	RIVET STEEL, PER CENT
Phosphorus	Bessemer.......	Not over 0.10	
	Open-hearth.. .	Not over 0.06	Not over 0.06
Sulphur....................		....	Not over 0.045

3. Ladle Analyses.—An analysis of each melt of steel shall be made by the manufacturer to determine the percentages of carbon, manganese, phosphorus, and sulfur. This analysis shall be made from a test ingot taken during the pouring of the melt. The chemical composition thus determined shall be reported to the purchaser or his representative, and shall conform to the requirements specified in Section 2.

4. Check Analyses.—Analyses may be made by the purchaser from finished material representing each melt. The phosphorus and sulfur content thus determined shall not exceed that specified in Section 2 by more than 25 per cent.

III. Physical Properties and Tests

5. Tension Tests.—(*a*) The material shall conform to the following requirements as to tensile properties:

Properties considered	Structural steel	Rivet steel
Tensile strength, pounds per square inch..................	55,000–65,000	46,000–56,000
Yield point, min., pounds per square inch..................	0.5 tensile strength	0.5 tensile strength
Elongation in 8 in., min., per cent	$\frac{1{,}400{,}000}{\text{Tensile strength}}$ [a]	$\frac{1{,}400{,}000}{\text{Tensile strength}}$
Elongation in 2 in., min., per cent	22	

(*b*) The yield point shall be determined by the drop of the beam of the testing machine.

6. Modifications in Elongation.—(*a*) For structural steel over ¾ in. in thickness, a deduction of 1 from the percentage of elongation in 8 in. specified in Section 5 (*a*) shall be made for each increase of ⅛ in. in thickness above ¾ in., to a minimum of 18 per cent.

(*b*) For structural steel under 5/16 in. in thickness, a deduction of 2.5 from the percentage of elongation in 8 in. specified in Section 5 (*a*) shall be made for each decrease of 1/16 in. in thickness below 5/16 in.

7. Bend Tests.—(*a*) The test specimen for plates, shapes, and bars, except as specified in Paragraphs (*b*) and (*c*), shall bend cold through 180 degrees without cracking on the outside of the bent portion, as follows: For material ¾ in. or under in thickness, flat on itself; for material over ¾ in. to and including 1¼ in. in thickness, around a pin the diameter of which is equal to the thickness of the specimen; and for material over 1¼ in. in thickness, around a pin the diameter of which is equal to twice the thickness of the specimen.

(*b*) The test specimen for pins, rollers, and other bars, when prepared as specified in Section 8 (*e*), shall bend cold through 180 degrees around a 1-in. pin without cracking on the outside of the bent portion.

[a] See Section 6.

(c) The test specimen for rivet steel shall bend cold through 180 degrees flat on itself without cracking on the outside of the bent portion.

8. Test Specimens.—(*a*, Tension and bend test specimens shall be taken from rolled steel in the condition in which it comes from the rolls, except as specified in Paragraph (*b*).

(*b*) Tension and bend test specimens for pins and rollers shall be taken from the finished bars, after annealing when annealing is specified.

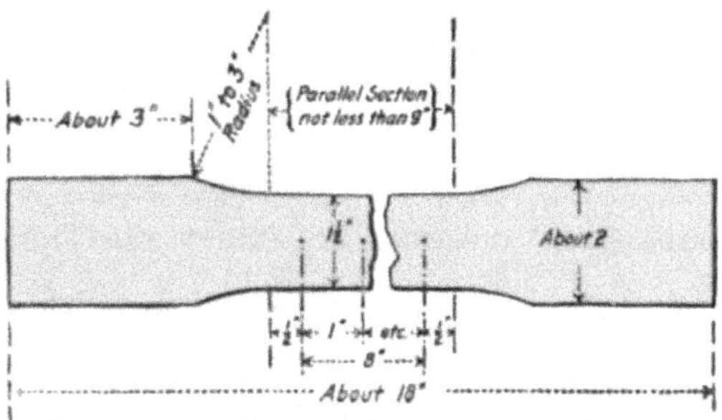

Fig. 1.—Tension and bend test specimen.

(*c*) Tension and bend test specimens for plates, shapes, and bars, except as specified in Paragraphs (*d*), (*e*), and (*f*), shall be of the full thickness of material as rolled; and may be machined to the form and dimensions shown in Fig. 1, or with both edges parallel.

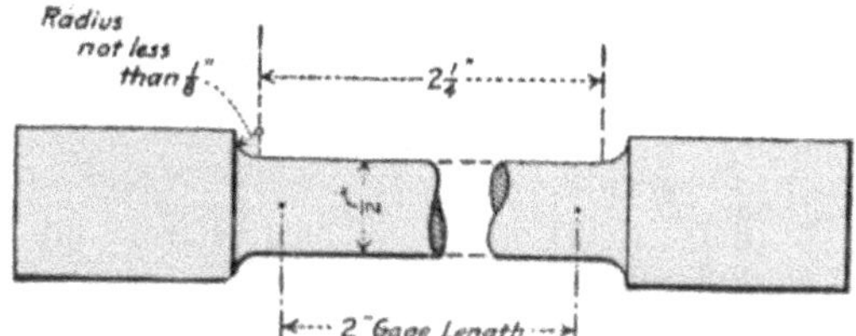

Note 1- The Gage Length, Parallel Portions and Fillets shall be as Shown, but the Ends may be of any Form which will Fit the Holders of the Testing Machine.

Fig. 2.—Tension test specimen.

(*d*) Tension and bend test specimens for plates over 1½ in. in thickness may be machined to a thickness or diameter of at least ¾ in. for a length of at least 9 in.

(*e*) Tension test specimens for pins, rollers, and bars over 1½ in. in thickness or diameter may conform to the dimensions shown in Fig. 2. In this case, the ends shall be of a form to fit the holders of the testing machine in such a way that the load shall be axial. Bend test specimens may be 1 by ½ in. in section. The axis of the specimen shall be located at any point midway between the center and surface and shall be parallel to the axis of the bar.

(*f*) Tension and bend test specimens for rivet steel shall be of the full-size section of bars as rolled.

9. Number of Tests.—(*a*) One tension and one bend test shall be made from each melt; except that if material from one melt differs ⅜ in. or more in thickness, one tension and one bend test shall be made from both the thickest and the thinnest material rolled.

(*b*) If any test specimen shows defective machining or develops flaws, it may be discarded and another specimen substituted.

(*c*) If the percentage of elongation of any tension test specimen is less than that specified in Section 5 (*a*) and any part of the fracture is more than ¾ in. from the center of the gage length of a 2-in. specimen or is outside the middle third of the gage length of an 8-in. specimen, as indicated by scribe scratches marked on the specimen before testing, a retest shall be allowed.

IV. Permissible Variations in Weight and Thickness

10. Permissible Variations.—The cross-section or weight of each piece of steel shall not vary more than 2.5 per cent from that specified; except in the case of sheared plates, which shall be covered by the following permissible variations. One cubic inch of rolled steel is assumed to weigh 0.2833 lb.

(*a*) *When Ordered to Weight per Square Foot.*—The weight of each lot[1] in each shipment shall not vary from the weight ordered more than the amount given in Table I.

(*b*) *When Ordered to Thickness.*—The thickness of each plate shall not vary more than 0.01 in. under that ordered.

The overweight of each lot[2] in each shipment shall not exceed the amount given in Table II.

V. Finish

11. Finish.—The finished material shall be free from injurious defects and shall have a workmanlike finish.

VI. Marking

12. Marking.—The name or brand of the manufacturer and the melt number shall be legibly stamped or rolled on all finished material, except that rivet and lattice bars and other small sections shall, when loaded for shipment, be properly separated and marked for identification. The identification marks shall be legibly stamped on the end of each pin and roller. The melt number shall be legibly marked, by stamping if practicable, on each test specimen.

VII. Inspection and Rejection

13. Inspection.—The inspector representing the purchaser shall have free entry, at all times while work on the contract of the purchaser is

[1] The term "lot" applied to Table I means all of the plates of each group width and group weight.

[2] The term "lot" applied to Table II means all of the plates of each group width and group thickness.

Table I.—Permissible Variations of Plates Ordered to Weight

Ordered weight, pounds per square foot	Permissible variations in average weights per square foot of plates for widths given, expressed in percentages of ordered weights																		Ordered weight, pounds per square foot
	Under 48 in.		48 to 60 in., excl.		60 to 72 in., excl.		72 to 84 in., excl.		84 to 96 in., excl.		96 to 108 in., excl.		108 to 120 in., excl.		120 to 132 in., excl.		132 in. or over		
	Over	Under	Over	Under	Over	Under	Over	Under	Over	Under	Over	Under	Over	Under	Over	Under	Over	Under	
Under 5.	5.0	3.[illegible]	5.5	3.0	6.0	3.0	7.0	3.0	...	...	...	...	...	...	...	...	...	...	Under 5
5.0 to 7.5 excl.	4.5	3.[illegible]	5.0	3.0	5.5	3.0	6.0	3.0	...	...	...	...	...	...	...	...	...	...	5.0 to 7.5 excl.
7.5 to 10.0 excl.	4.0	3.[illegible]	4.5	3.0	5.0	3.0	5.5	3.0	6.0	3.0	7.0	3.0	8.0	3.0	...	...	...	...	7.5 to 10.0 excl.
10.0 to 12.5 excl.	3.5	2.5	4.0	3.0	4.5	3.0	5.0	3.0	5.5	3.0	6.0	3.0	7.0	3.0	8.0	3.0	9.0	3.0	10.0 to 12.5 excl.
12.5 to 15.0 excl.	3.0	2.5	3.5	2.5	4.0	3.0	4.5	3.0	5.0	3.0	5.5	3.0	6.0	3.0	7.0	3.0	8.0	3.0	12.5 to 15.0 excl.
15.0 to 17.5 excl.	2.5	2.5	3.0	2.5	3.5	2.5	4.0	3.0	4.5	3.0	5.0	3.0	5.5	3.0	6.0	3.0	7.0	3.0	15.0 to 17.5 excl.
17.5 to 20.0 excl.	2.5	2.0	2.5	2.5	3.0	2.5	3.5	2.5	4.0	3.0	4.5	3.0	5.0	3.0	5.5	3.0	6.0	3.0	17.5 to 20.0 excl.
20.0 to 25.0 excl.	2.0	2.0	2.5	2.0	2.5	2.5	3.0	2.5	3.5	2.5	4.0	3.0	4.5	3.0	5.0	3.0	5.5	3.0	20.0 to 25.0 excl.
25.0 to 30.0 excl.	2.0	2.0	2.0	2.0	2.5	2.0	2.5	2.5	3.0	2.5	3.5	3.0	4.0	3.0	4.5	3.0	5.0	3.0	25.0 to 30.0 excl.
30.0 to 40.0 excl.	2.0	2.0	2.0	2.0	2.0	2.0	2.5	2.0	2.5	2.5	3.0	2.5	3.5	3.0	4.0	3.0	4.5	3.0	30.0 to 40.0 excl.
40 or over.	2.0	2.0	2.0	2.0	2.0	2.0	2.0	2.0	2.5	2.0	2.5	2.5	3.0	2.5	3.5	3.0	4.0	3.0	40 or over

Note.—The weight per square foot of individual plates shall not vary from the ordered weight by more than 1½ times the amount given in this table.

Table II.—Permissible Overweights of Plates Ordered to Thickness

Ordered thickness, inches	Permissible excess in average weights per square foot of plates for widths given, expressed in percentages of nominal weights									Ordered thickness, inches
	Under 48 in.	48 to 60 in., excl.	60 to 72 in., excl.	72 to 84 in., excl.	84 to 96 in., excl.	96 to 108 in., excl.	108 to 120 in., excl.	120 to 132 in., excl.	132 in. or over	
Under 1/8.	9.0	10.0	12.0	14.0	....	....	....	....	....	Under 1/8
1/8 to 3/16 excl.	8.0	9.0	10.0	12.0	....	....	....	....	....	1/8 to 3/16 excl.
3/16 to 1/4 excl.	7.0	8.0	9.0	10.0	12.0	....	....	....	....	3/16 to 1/4 excl.
1/4 to 5/16 excl.	6.0	7.0	8.0	9.0	10.0	12.0	14.0	16.0	19.0	1/4 to 5/16 excl.
5/16 to 3/8 excl.	5.0	6.0	7.0	8.0	9.0	10.0	12.0	14.0	17.0	5/16 to 3/8 excl.
3/8 to 7/16 excl.	4.5	5.0	6.0	7.0	8.0	9.0	10.0	12.0	15.0	3/8 to 7/16 excl.
7/16 to 1/2 excl.	4.0	4.5	5.0	6.0	7.0	8.0	9.0	10.0	13.0	7/16 to 1/2 excl.
1/2 to 5/8 excl.	3.5	4.0	4.5	5.0	6.0	7.0	8.0	9.0	11.0	1/2 to 5/8 excl.

being performed, to all parts of the manufacturers' works which concern the manufacture of the material ordered. The manufacturer shall afford the inspector, free of cost, all reasonable facilities to satisfy him that the material is being furnished in accordance with these specifications. All tests (except check analyses) and inspection shall be made at the place of manufacture prior to shipment, unless otherwise specified, and shall be so conducted as not to interfere unnecessarily with the operation of the works.

14. Rejection.—(*a*) Unless otherwise specified, any rejection based on tests made in accordance with Section 4 shall be reported within five working days from the receipt of samples.

(*b*) Material which shows injurious defects subsequent to its acceptance at the manufacturers' works will be rejected, and the manufacturer shall be notified.

15. Rehearing.—Samples tested in accordance with Section 4, which represent rejected material, shall be preserved for two weeks from the date of the test report. In case of dissatisfaction with the results of the tests, the manufacturer may make claim for a rehearing within that time.

APPENDIX C

List of Standards and Tentative Standards of the American Society for Testing Materials

A. Ferrous Metals

A 1–14. Standard Specifications for Carbon-steel Rails.

A 2–12. Standard Specifications for Open-hearth Steel Girder and High Tee Rails.

A 3–14. Standard Specifications for Low-carbon-steel Splice Bars.

A 4–14. Standard Specifications for Medium-carbon-steel Splice Bars.

A 5–14. Standard Specifications for High-carbon-steel Splice Bars.

A 6–14. Standard Specifications for Extra-high-carbon-steel Splice Bars.

A 7–16. Standard Specifications for Structural Steel for Bridges.

A 8–16. Standard Specifications for Structural Nickel Steel.

A 9–16. Standard Specifications for Structural Steel for Buildings.

A 10–16. Standard Specifications for Structural Steel for Locomotives.

A 11–16. Standard Specifications for Structural Steel for Cars.

A 12–16. Standard Specifications for Structural Steel for Ships.

A 13–14. Standard Specifications for Rivet Steel for Ships.

A 14–16. Standard Specifications for Carbon-steel Bars for Railway Springs.

A 15–14. Standard Specifications for Billet-steel Concrete Reinforcement Bars.

A 16–14. Standard Specifications for Rail-steel Concrete Reinforcement Bars.

A 17–18. Standard Specifications for Carbon-steel and Alloy-steel Blooms, Billets, and Slabs for Forgings.

A 18–18. Standard Specifications for Carbon-steel and Alloy-steel Forgings.

A 19–18. Standard Specifications for Quenched and Tempered Carbon-steel Axles, Shafts, and Other Forgings for Locomotives and Cars.

A 20–16. Standard Specifications for Carbon-steel Forgings for Locomotives.

A 21–18. Standard Specifications for Carbon-steel Car and Tender Axles.

A 22–16. Standard Specifications for Cold-rolled Steel Axles.

A 23. Discontinued—Replaced by Specification A 57.

A 24. Discontinued—Replaced by Specification A 57.

A 25–16. Standard Specifications for Wrought, Solid Carbon-steel Wheels for Electric Railway Service.

A 26–16. Standard Specifications for Steel Tires.

A 27-16. Standard Specifications for Steel Castings.
A 28-18. Standard Specifications for Lap-welded and Seamless Steel Boiler Tubes for Locomotives.
A 29-18. Standard Specifications for Automobile Carbon and Alloy Steels.
A 30-18. Standard Specifications for Boiler and Firebox Steel for Locomotives.
A 31-14. Standard Specifications for Boiler-rivet Steel.
A 32-14. Standard Specifications for Cold-drawn Bessemer Steel Automatic Screw Stock.
A 33-14. Standard Methods for Chemical Analysis of Plain Carbon Steel.
A 34-18. Standard Tests for Magnetic Properties of Iron and Steel.
A 35-11. Recommended Practice for Annealing of Miscellaneous Rolled and Forged Carbon-steel Objects.
A 36-14. Recommended Practice for Annealing of Carbon-steel Castings.
A 37-14. Recommended Practice for Heat Treatment of Case-hardened Carbon-steel Objects.
A 38-18. Standard Specifications for Lap-welded Charcoal-iron Boiler Tubes for Locomotives.
A 39-18. Standard Specifications for Staybolt Iron.
A 40-18. Standard Specifications for Engine-bolt Iron.
A 41-18. Standard Specifications for Refined Wrought-iron Bars.
A 42-18. Standard Specifications for Wrought-iron Plates.
A 43-09. Standard Specifications for Foundry Pig Iron.
A 44-04. Standard Specifications for Cast-iron Pipe and Special Castings.
A 45-14. Standard Specifications for Cast-iron Locomotive Cylinders.
A 46-05. Standard Specifications for Cast-iron Car Wheels.
A 47-19. Standard Specifications for Malleable Castings.
A 48-18. Standard Specifications for Gray-iron Castings.
A 49-15. Standard Specifications for Quenched High-carbon-steel Splice Bars.
A 50-16. Standard Specifications for Quenched Carbon-steel Track Bolts.
A 51-16. Standard Specifications for Quenched Alloy-steel Track Bolts.
A 52-18. Standard Specifications for Lap-welded and Seamless Steel and Wrought-iron Boiler Tubes for Stationary Service.
A 53-18. Standard Specifications for Welded Steel Pipe.
A 53-20 T. Tentative Specifications for Welded Steel Pipe.
A 54-15. Standard Specifications for Cold-drawn Open-hearth Steel Automatic Screw Stock.
A 55-15. Standard Methods for Chemical Analysis of Alloy Steels.
A 56-18. Standard Specifications for Iron and Steel Chain.
A 57-16. Standard Specifications for Wrought, Solid Carbon-steel Wheels for Steam Railway Service.
A 58-16. Standard Specifications for Carbon-steel Bars for Vehicle and Automobile Springs.
A 59-16. Standard Specifications for Silico-manganese-steel Bars for Automobile and Railway Springs.
A 60-16. Standard Specifications for Chrome-vanadium-steel Bars for Automobile and Railway Springs.

A 61–16. Standard Specifications for Helical Steel Springs for Railways.
A 62–16. Standard Specifications for Elliptical Steel Springs for Railways.
A 63–18. Standard Specifications for Quenched-and-tempered Alloy-steel Axles, Shafts, and Other Forgings for Locomotives and Cars.
A 64–16. Standard Methods for Sampling and Analysis of Pig and Cast Iron.
A 65–18. Standard Specifications for Steel Track Spikes.
A 66–18. Standard Specifications for Steel Screw Spikes.
A 67–20 T. Tentative Specifications for Steel Tie Plates.
A 68–18. Standard Specifications for Carbon-steel Bars for Railway Springs with Special Silicon Requirements.
A 69–18. Standard Specifications for Elliptical Steel Springs for Automobiles.
A 70–18 T. Tentative Specifications for Boiler and Firebox Steel for Stationary Service.
A 71–20 T. Tentative Specifications for Carbon Tool Steel.
A 72–18. Standard Specifications for Welded Wrought-iron Pipe.
A 73–18. Standard Specifications for Wrought-iron Rolled or Forged Blooms and Forgings for Locomotives and Cars.
A 74–18. Standard Specifications for Cast-iron Soil Pipe and Fittings.
A 75. Discontinued—Has become A 47.
A 76–20 T. Tentative Specifications for Low-carbon-steel Track Bolts.
A 77–20 T. Tentative Specifications for Electric Cast-steel Anchor Chain.
A 78–20 T. Tentative Specifications for Steel Plates for Forge Welding.
A 79–19 T. Tentative Specifications for Extra Refined Wrought-iron Bars.
A 80–20 T. Tentative Specifications for Commercial Bar Steels.
A 81–20 T. Tentative Definitions of Terms Relating to Wrought-iron Specifications.

B. Non-ferrous Metals

B 1–15. Standard Specifications for Hard-drawn Copper Wire.
B 2–15. Standard Specifications for Medium Hard-drawn Copper Wire.
B 3–15. Standard Specifications for Soft or Annealed Copper Wire.
B 4–13. Standard Specifications for Lake Copper Wire Bars, Cakes, Slabs, Billets, Ingots, and Ingot Bars.
B 5–13. Standard Specifications for Electrolytic Copper Wire Bars, Cakes, Slabs, Billets, Ingots, and Ingot Bars.
B 6–18. Standard Specifications for Spelter.
B 7–14. Standard Specifications for Manganese-bronze Ingots for Sand Castings.
B 8–16. Standard Specifications for Bare Concentric-lay Copper Cable: Hard, Medium-hard, or Soft.
B 9–16. Standard Specifications for High-strength Bronze Trolley Wire, Round and Grooved: 40 and 65 per cent Conductivity.
B 10–18. Standard Specifications for the Alloy: Copper, 88 per cent; Tin, 10 per cent; Zinc, 2 per cent.
B 11–18. Standard Specifications for Copper Plates for Locomotive Fireboxes.

B 12–18. Standard Specifications for Copper Bars for Locomotive Staybolts.
B 13–18. Standard Specifications for Seamless Copper Boiler Tubes.
B 14–18. Standard Specifications for Seamless Brass Boiler Tubes.
B 15–18. Standard Specifications for Brass Forging Rod.
B 16–18. Standard Specifications for Free-cutting Brass Rod for Use in Screw Machines.
B 17–18 T. Tentative Specifications for Non-ferrous Alloys for Railway Equipment in Ingots, Castings, and Finished Car and Tender Bearings.
B 18–20 T. Tentative Methods for Chemical Analysis of Alloys of Lead, Tin, Antimony, and Copper.
B 19–19. Standard Specifications for Cartridge Brass.
B 20–19. Standard Specifications for Cartridge Brass Disks.
B 21–19 Standard Specifications for Naval Brass Rods for Structural Purposes.
B 22–18 T. Tentative Specifications for Bronze Bearing Metals for Turntables and Movable Railroad Bridges.
B 23–18 T. Tentative Specifications for White Metal Bearing Alloys (known commercially as "Babbitt Metal").
B 24–20 T. Tentative Specifications for Aluminum Ingots for Remelting and for Rolling.
B 25–19 T. Tentative Specifications for Aluminum Sheet.
B 26–19 T. Tentative Specifications for Light Aluminum Casting Alloys.
B 27–19. Standard Methods for Chemical Analysis of Manganese Bronze.
B 28–19. Standard Methods for Chemical Analysis of Gun Metal.
B 29–20 T. Tentative Specifications for Pig Lead.
B 30–19 T. Tentative Specifications for Brass Ingot Metal for Sand Castings.
B 31–19 T. Tentative Specifications for Bronze Bearing Metal in Ingot Form.
B 32–19 T. Tentative Specifications for Solder Metal.
B 33–19 T. Tentative Specifications for Tinned Soft or Annealed Copper Wire for Rubber Insulation.
B 34–20. Standard Methods for Battery Assay of Copper.
B 35–20. Standard Methods for Chemical Analysis of Pig Lead.
B 36–20 T. Tentative Specifications for Sheet High Brass.
B 37–20 T. Tentative Specifications for Aluminum for Use in the Manufacture of Iron and Steel.

C. Cement, Lime, Gypsum, and Clay Products

C 1. Discontinued—Replaced by Specifications C 9 and C 10.
C 2. Discontinued—Replaced by Specification C 19.
C 3. Discontinued—Replaced by Specification C 19.
C 4–16. Standard Specifications for Drain Tile.
C 5–15. Standard Specifications for Quicklime.
C 6–15. Standard Specifications for Hydrated Lime.

C 6–19 T. Tentative Specifications for Masons' Hydrated Lime.
C 7–15. Standard Specifications for Paving Brick.
C 8–15. Standard Definitions of Terms Relating to Sewer Pipe.
C 9–21. Standard Specifications and Tests for Portland Cement.
C 9–16 T. Tentative Specifications and Tests for Compressive Strength of Portland-cement Mortars.
C 10–09. Standard Specifications for Natural Cement.
C 11–16 T. Tentative Definitions of Terms Relating to the Gypsum Industry.
C 12–19. Recommended Practice for Laying Sewer Pipe.
C 13–20. Standard Specifications for Clay Sewer Pipe.
C 14–20. Standard Specifications for Cement-concrete Sewer Pipe.
C 15–17 T. Tentative Specifications for Required Safe Crushing Strengths of Sewer Pipe to Carry Loads from Ditch Filling.
C 16–20. Standard Test for Refractory Materials under Load at High Temperatures.
C 17–19 T. Tentative Test for Slagging Action of Refractory Materials.
C 18–20. Standard Methods for Ultimate Chemical Analysis of Refractory Materials.
C 18–20 T. Tentative Method for Ultimate Chemical Analysis of Chrome Ores and Chrome Brick.
C 19–18. Standard Specifications for Fire Tests of Materials and Construction.
C 20–20. Standard Test for Porosity and Permanent Volume Changes in Refractory Materials.
C 21–20. Standard Specifications for Building Brick.
C 22–20 T. Tentative Specifications for Gypsum.
C 23–20 T. Tentative Specifications for Calcined Gypsum.
C 24–20. Standard Test for Softening Point of Fireclay Brick.
C 25–19 T. Tentative Methods for Chemical Analysis of Limestone, Lime, and Hydrated Lime.
C 26–20 T. Tentative Methods for Tests of Gypsum and Gypsum Products.
C 27–20. Standard Definitions for Clay Refractories.
C 28–20 T. Tentative Specifications for Gypsum Plasters.
C 29–20 T. Tentative Test for Unit Weight of Aggregate for Cement Concrete.
C 30–20 T. Tentative Method for Determination of Voids in Fine Aggregate for Cement Concrete.
C 31–20 T. Tentative Methods for Making and Storing Specimens of Concrete in the Field.

D. Miscellaneous Materials

D 1–15. Standard Specifications for Purity of Raw Linseed Oil from North American Seed.
D 2–08. Standard Test for Abrasion of Road Material.
D 3–18. Standard Test for Toughness of Rock.
D 4–11. Standard Test for Soluble Bitumen.

D 5–16. Standard Test for Penetration of Bituminous Materials.
D 6–20. Standard Test for Loss on Heating of Oil and Asphaltic Compounds.
D 7–18. Standard Method for Making a Mechanical Analysis of Sand or Other Fine Highway Material, Except Fine Aggregates Used in Cement Concrete.
D 8–18. Standard Definitions of Terms Relating to Materials for Roads and Pavements.
D 9–15. Standard Definitions of Terms Relating to Structural Timber.
D 10–15. Standard Specifications for Yellow-pine Bridge and Trestle Timbers.
D 11–15. Standard Specifications for Purity of Boiled Linseed Oil from North American Seed.
D 12–16. Standard Specifications for Purity of Raw Tung Oil.
D 13–15. Standard Specifications for Turpentine.
D 13–20 T. Tentative Specifications for Turpentine.
D 14–15. Standard Specifications for 2⅝-in. Cotton Rubber-lined Fire Hose for Private Department Use.
D 15–15. Standard Methods for Testing of Cotton Rubber-lined Hose.
D 16–15. Standard Definitions of Terms Relating to Paint Specifications.
D 17–16. Standard Specifications for Foundry Coke.
D 18–16. Standard Method for Making a Mechanical Analysis of Broken Stone or Broken Slag, except Aggregates Used in Cement Concrete.
D 19–16. Standard Method for Making a Mechanical Analysis of Mixtures of Sand or Other Fine Material with Broken Stone or Broken Slag, except Aggregates Used in Cement Concrete.
D 20–18. Standard Method for Distillation of Bituminous Materials Suitable for Road Treatment.
D 21–16. Standard Methods for Sampling of Coal.
D 22–16. Standard Methods for Laboratory Sampling and Analysis of Coal.
D 22–19 T. Tentative Method for Determination of Fusibility of Coal Ash.
D 23–20 T. Tentative Specifications for Structural Douglas Fir.
D 24–20. Standard Specifications for Southern Yellow-pine Timber to be Creosoted.
D 25–20. Standard Specifications for Southern Yellow-pine Piles and Poles to be Creosoted.
D 26–18. Standard Specifications for 2½-, 3-, and 3½-in. Double-jacketed Cotton Rubber-lined Fire Hose for Public Fire Department Use.
D 27–16 T. Tentative Specifications for Insulated Wire and Cable: 30-Per Cent Hevea Rubber.
D 28–17. Standard Tests for Paint Thinners Other than Turpentine.
D 29–17. Standard Tests for Shellac.
D 30–18. Standard Test for Determination of Apparent Specific Gravity of Coarse Aggregates.
D 31. Discontinued—Replaced by Method D 39.
D 32. Discontinued—Replaced by Method D 39.

D 33. Discontinued—Replaced by Method D 39.
D 34–17. Standard Methods for Routine Analysis of White Pigments.
D 35–18. Standard Form of Specifications for Certain Commercial Grades of Broken Stone.
D 36–19. Standard Method for Determination of Softening Point of Bituminous Materials Other than Tar Products (Ring-and-Ball Method).
D 37–18. Standard Methods for Laboratory Sampling and Analysis of Coke.
D 38–18. Standard Methods for Sampling and Analysis of Creosote Oil.
D 39–20. Standard General Methods for Testing Cotton Fabrics.
D 40–17 T. Tentative Specifications for Asphalt for Use in Damp-proofing and Waterproofing.
D 41–17 T. Tentative Specifications for Primer for Use with Asphalt for Use in Damp-proofing and Waterproofing.
D 42–17 T. Tentative Specifications for Coal-tar Pitch for Use in Damp-proofing and Waterproofing.
D 43–17 T. Tentative Specifications for Creosote Oil for Priming Coat with Coal-tar Pitch for Use in Damp-proofing and Waterproofing.
D 44–20 T. Tentative Specifications for Wooden Boxes, Nailed and Lock-corner Construction, for the Shipment of Canned Foods.
D 45–17 T. Tentative Specifications for Canned Foods Boxes, Wire-bound Construction.
D 46–18. Standard Specifications for Air-line Hose for Pneumatic Tools.
D 47–18. Standard Tests for Lubricants.
D 47–20. Standard Test for Viscosity of Lubricants.
D 48–17 T. Tentative Tests for Molded Insulating Materials.
D 49–18. Standard Methods for Routine Analysis of Dry Red Lead.
D 50–18. Standard Methods for Routine Analysis of Yellow, Orange, Red, and Brown Pigments containing Iron and Manganese.
D 51–18 T. Tentative Specifications for Foots Permissible in Properly Clarified Pure Raw Linseed Oil from North American Seed.
D 52–20. Standard Specifications for Wooden Paving Blocks for Exposed Pavements.
D 53–20. Standard Specifications for Rubber Belting for Power Transmission.
D 54–20. Standard Specifications for Steam Hose.
D 55–19. Standard Tests for Determination of Apparent Specific Gravity of Sand, Stone, and Slag Screenings and Other Fine Non-bituminous Highway Materials.
D 56–19. Standard Test for Flash Point of Volatile Paint Thinners.
D 57–20. Standard Specifications for Materials for Cement Grout Filler for Brick and Stone-block Pavements.
D 58–20. Standard Specifications for Materials for Cement Mortar Bed for Brick, Stone-block, and Wood-block Pavements.
D 59–19 T. Tentative Specifications for Block for Granite-block Pavements.
D 60–20. Standard Specifications for Leader Hose for Use with Pneumatic Tools.

D 61–20. Standard Method for Determination of Softening Point of Tar Products (Cube-in-water Method).

D 62–20 T. Tentative Specifications for Workability of Concrete for Concrete Pavements.

D 63–20 T. Tentative Specifications for Commercial Sizes of Broken Stone and Broken Slag for Highway Construction.

D 64–20 T. Tentative Specifications for Commercial Sizes of Sand and Gravel for Highway Construction.

D 65–20 T. Tentative Specifications for Broken Slag for Waterbound Base.

D 66–20 T. Tentative Specifications for Shovel-run or Crusher-run Broken Slag for Waterbound Base.

D 67–20 T. Tentative Specifications for Natural or Artificial Sand-clay Mixtures for Highway Surfacing.

D 68–20 T. Tentative General Specifications for Wooden Boxes, Nailed and Lock-corner Construction.

D 69–20 T. Tentative Specifications for Adhesive Insulating Tape.

D 70–20 T. Tentative Test for Specific Gravity of Road Oils, Road Tars, Asphalt Cements, and Soft Tar Pitches.

D 71–20 T. Tentative Test for Specific Gravity of Asphalts and Tar Pitches Sufficiently Solid to be Handled in Fragments.

D 72–20 T. Tentative Test for Quantity of Clay and Silt in Gravel for Highway Construction.

D 73–20 T. Tentative Test for Quantity of Clay in Sand-clay, Topsoil, and Semi-gravel for Highway Construction.

D 74–20 T. Tentative Test for Quantity of Clay and Silt in Sand for Highway Construction.

D 75–20 T. Tentative Methods for Sampling of Stone, Slag, Gravel, Sand, and Stone Block for Use as Highway Materials, Including Some Material Survey Methods.

D 76–20 T. Tentative Methods for Testing Textiles.

E. Miscellaneous Subjects

E 1–18. Standard Methods for Testing.

E 2–20. Standard Definitions and Rules Governing the Preparation of Micrographs of Metals and Alloys.

INDEX

www.ingramcontent.com/pod-product-compliance
Lightning Source LLC
LaVergne TN
LVHW021939220826
846092LV00010B/1176

* 9 7 8 9 3 5 5 2 8 2 2 2 4 *